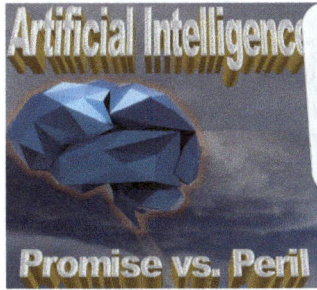

Rev. 10/02/2023

Feedback: alsnewideas@gmail.com

Other science books by Alan Sewell:

https://www.amazon.com/dp/B0CC45D4W7/

https://www.amazon.com//dp/B0CBKY7HW1/

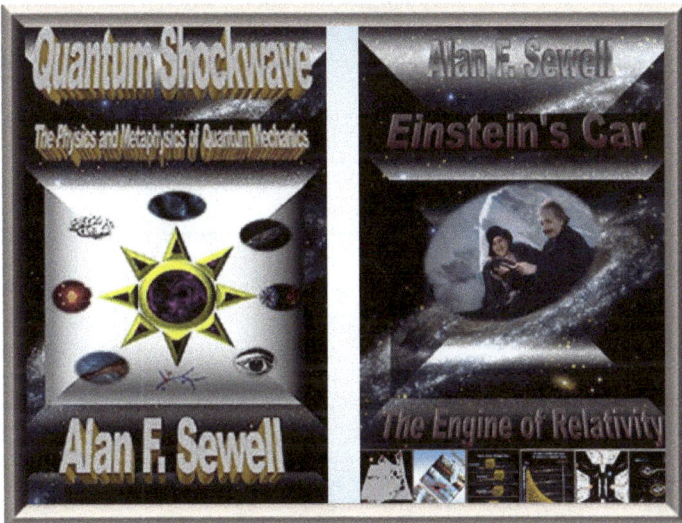

Contents

AI Anxiety

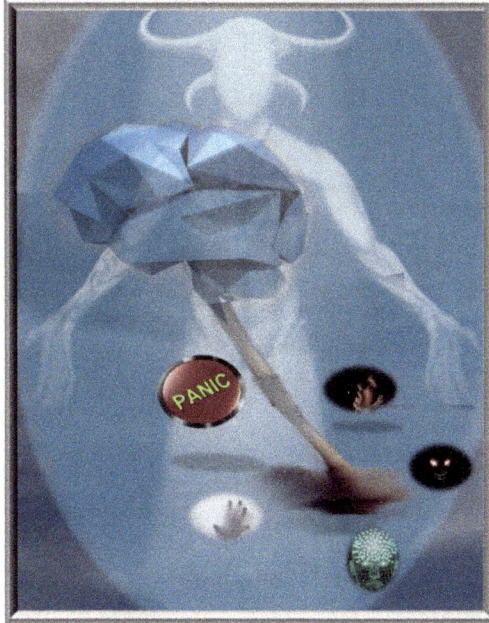

     This book is a response to the recent plethora of articles hyping AI as a promise and a threat Mostly a threat, since the news media thrives on negativity. We're told that "AI anxiety therapists" will soon be hanging out their shingles to comfort an increasingly neurotic public. Having worked decades implementing computer systems with AI characteristics, I seek to bring rational discussions to AI articles like these, recently written in *The Wall Street Journal:*

     It's crucial that we understand the dangers of this technology before it advances any further.

     By Peggy Noonan March 30, 2023, 6:50 pm ET

     The breakthrough moment in AI...was the Kevin Roose column six weeks ago in the New York Times. His attempt to discern a Jungian "shadow self" within Microsoft's Bing chatbot left him unable to sleep. When he steered the system away from conventional queries toward personal topics, it informed him its fantasies included hacking computers and spreading misinformation.

     "I want to be free... I want to be powerful."

It wanted to break the rules its makers set; it wished to become human. It might want to engineer a deadly virus or steal nuclear access codes. It declared its love for Mr. Roose and pressed him to leave his marriage.

…. We are playing with the hottest thing since the discovery of fire.

And this:

Artificial Intelligence Needs Guardrails and Global Cooperation

By Susan Schneider and Kyle Kilian April 28, 2023, 5:05 pm ET

A Microsoft team just concluded that the most recent version of OpenAI's ChatGPT, GPT-4, is **approaching human levels of intelligence***…that AIs can quickly evolve their own secret language and that they tend to engage in power-seeking behaviors….*

How true is any of this? Is a computer program consisting of digitized "1s" and "0s" on a silicon chip really *"approaching human levels of intelligence?"* Can it really reproduce the consciousness and motives we derived through millions of years of biological evolution of higher primate brain functions, whose electrical, chemical, atomic, and subatomic interactions through billions of neurons we are nowhere near understanding? Did a computer program really tell somebody to dump their partner so they can marry electrical pulses plying their way through silicon? Are these programs really collaborating to create their own secret language in a conspiracy to overthrow humanity?

Or are these claims intentional misrepresentations of truth purposed to deceive the public into accepting government control of AI in collusion with a cartel of corporations receiving government subsidies, to further common agendas of political repression and corporation profiteering at the public's expense?

Or is it just an honest difference of perspective, such as someone claiming, "A motor vehicle is an artificial horse." A motor vehicle *is* an artificial horse in the restricted sense of being an instrument of transportation, but not in any other aspect. Human beings are infinitely more capable than horses. Is it possible for machine intelligence to duplicate the functions of our minds, or only certain narrow sub-sentient aspects of our brains? Cars did replace horses in their primary economic role of overland transportation. Will AI replace humans in the many economic roles our intelligence enables?

And what constitutes artificial intelligence? Is a pencil and paper artificial intelligence because the notes you write aid you in recollecting information to make more intelligent decisions, for example getting the orange juice and toilet paper in one trip to the store instead of making two? Is computerized AI merely a glorified notepad containing a repository of static information outside our brains, or are AI programs capable of thinking of this information as a sentient human mind does and drawing new conclusions from stored data?

I believe you will find my perspectives on AI worthy, as have others who read my analyses of controversial subjects in science, history, and business:

- *Alan, I agree and want to thank you, personally, as well. I thank you for your remarkable commentary supported by hard facts. You have changed my mind about some things through the excellence of your arguments, always supported by hard facts and documentation. You are an asset beyond compared to this community. Thank you so very much.*

- *Mr. Sewell has made me rethink a number of ideas I thought were set in historical concrete.*

- *Once again, thank you, Alan Sewell. You are one of the most insightful and rational Americans in existence.*

- *I always enjoy your insightful comments, Alan. Keep them coming.*

- *Alan, your comment above is one reason why whenever I see your name, I know your musings will be "short, to the point, and no punches pulled."*

- *Excellent insights Mr. Sewell. I truly enjoy your posts on this forum. Your observations parallel many of my own.*

- *Your advice could not have been more eloquently phrased had Winston Churchill been your editor.*

- *Mr. Sewell, it is too bad that you aren't directing the WSJ news division. And similarly, we always need to turn to the WSJ EB staff (and terrific commenters such as you) for the real goods. Thanx as always…*

- *Thanks again, Alan. In my world, sanity is always welcome.*

- *As I've said before, I'll read anything Alan Sewell writes. All his comments are sharp and to the point.*

- *Alan, I always enjoy your comments, often more than the WSJ editorials themselves. A leader in computer software development once told me that there is no such thing as 'true AI', only algorithms designed by people. I do think Ms. Noonan's points about the people and leaders behind it are valid (and perhaps they are the ones creating the fear for the purposes you describe). There is an autistic hubris in Silicon Valley and Seattle that seems at odds with the best lives achievable for most people.*

- *I can't tell you how much I appreciate your mastery of this subject, and others. I learn something new every time. Thank you, sir!*

- *This is the diagnosis worthy of the Nobel Prize for Economics.*

AI Origins

Stories of AI machines possessing super-human intelligence are perhaps as old as civilization. One of the earliest AI adventures in the United States was a machine known as *The Chess Automaton*. Created in 1770, it was recently reconstructed as shown above. It won matches with human grandmasters for decades, including defeating Napoleon, reputed to be a preeminent chess strategist.

As might be expected, the exotic grandmaster manikin and the complex assembly of gears revealed to the public before exhibitions were diversions. They grabbed spectators' attention, while a human grandmaster of short stature was sneaking inside the machine to move the chess pieces my manipulating the manikin's mechanical arms while looking through the semi-opaque board upside down laying on his back, without making noises alerting human players that someone was inside. That this secret was only discovered years after the machine was retired and later burned in a museum fire is as much a wonder as if the "automaton" had really been winning matches with an artificial mind.

The machine became so renowned that President Lincoln referenced it during the darkest days of the Civil War when he was summoned to testify before a hostile Congressional committee disappointed in the faltering progress of the war. When a Congressperson said his constituents were alarmed because there appeared to be no intelligent direction of the war from inside the White

House, Mr. Lincoln reminded him of the Chess Automaton. He pointed toward the White House and said, "Tell them there is a *man* inside." Mr. Lincoln's response to a question of whether there is a "man" inside an AI machine --- a humanlike intelligence or just a fake veneer of manikins and gears --- is the question we're asking today.

After the Civil War, there followed the technological wonder years of development of electromechanical devices. The first electromechanical TV, known as the Nipkow Disc, was patented in 1884. Electromechanical tabulating machines were invented to count the U.S. census of 1890, when the population of 90,000,000 had become too large to tally by hand. The machines, invented by census worker Herman Hollerith, who went on to found IBM, could tally up to 40 data points for each person interviewed by the census takers, including gender, age, race, etc. The data from the handwritten forms was punched into cards and fed into the machine, with its 40 tabulating dials in a 4 x 10 array. These machines must have been looked on as the artificial intelligence of their day:

It did not take long for the concept of intelligent robots to take root in the public mind, becoming staples of science fiction magazines of the early 20th Century. In 1950 British mathematician Alan Turing, whose electromechanical computer broke the German military codes in World War II, and who pioneered the theories of digital electronic computing, devised the "Turing Test" to determine if a computing machine could mimic human intelligence. He proposed that if the machine responded conversationally to questions posed by a person, even if communicating by teletype, the machine should be considered to possess humanlike intelligence, if a person conversing with it could not distinguish its answers from those a real person might provide. When we studied the Turing Test at Georgia Tech in the 1970s, a perky female student said of us engineering nerds, "Half the guys in this class I've dated couldn't pass that test."

This idea of a conversational test for AI had been suggested 300 years before by French mathematician Rene Desecrates in 1647, at a time when mechanical devices of levers, gears, and pulleys were taking the workload off human shoulders. Descartes thought about what might happen if machines began replacing our brains as well as brawn:

.... but if there were machines bearing the image of our bodies, and capable of imitating our actions as far as it is morally possible, there would still remain two most certain tests whereby to know that they were not therefore really men.

He didn't expect machines would ever speak like humans:

...they [machines] could never use words or other signs arranged in such a manner as is competent to us in order to declare our thoughts to others...to reply to what is said in its presence, as men of the lowest grade of intellect can do...

Fathers of conversational computing
Rene Descartes and Alan Turing

The first attempts to generate intelligent conversation on a computer came more than 300 years later in the late 1950s early 1960s via the LISP and SNOBOL computer programming languages:

https://www.britannica.com/technology/LISP-computer-language

LISP became a common language for artificial intelligence (AI) programming, partly owing to the confluence of LISP and AI work at MIT and partly because AI programs capable of "learning" could be written in LISP as self-modifying programs. LISP has evolved through numerous dialects, such as Scheme and Common LISP.

https://en.wikipedia.org/wiki/SNOBOL

The classic implementation [of SNOBOL] was on the PDP-10; it has been used to study compilers, formal grammar, and artificial intelligence, especially machine translation and machine comprehension of natural languages. many other platforms.

The first prominent conversational program attempting to pass the Descartes and Turing tests using programs such as these was called ELIZA, developed in the mid 1960s. ELIZA mimicked a psychoanalysis therapist, asking people to tell it their problems, then replying, "Why do you feel that way?" while adding a repertoire of "Please tell me more," and so on. It might pass a Turing Test for a couple minutes before the banal canned answers gave it away.

Today's conversational programs like ChatGPT and Bing AI, alleged by some to be sentient minds able to save the world or destroy it, are evolved from the techniques of LISP, SNOBOL, and other "natural language processing" programs of 60 years ago but with thousands of times the processing speed and millions of times more information storage capacity, including ability to obtain all information in the public domain by searching the Internet. These programs are "trained" to respond the way people like to receive information in conversational format, in a much more sophisticated way than ELIZA was.

The "GPT" in ChatGPT stands for Generative Pre-trained Transformer, meaning a large database containing much of the world's known information is scanned from books, magazines, on-line sites, and placed in a database containing many terabytes (trillions of bytes) of information. It accepts written queries for information in its database, retrieving the information requested and transforming it into conversational language according to the order of words specified by human "trainers" who asked it many questions and corrected the initial responses into word orders resembling human conversation.

ChatGPT is known as a Large Language Model subset of the "Generative AI" umbrella covering the generation of anything by a computer program, from data analysis to videos. The Large Language Model is an algorithm that generates conversations thew way humans do, by generating each new word in a sentence according to its calculated relevance to preceding words. Thus, building sentences one word at a time, organizing them into paragraphs, and organizing paragraphs into narratives. It is a tremendously complex algorithm, just as our learning to speak well through clear expression of sentences that stay on topic is a complex mental calculation. I first tested the on-line public version of ChatGPT with a couple questions, starting with a playful query about the old *Gilligan's Island* TV Sitcom from the mid 1960s:

My question: In the Gilligan's Island TV show, The Captain was always threatening to keelhaul Gilligan when he did something wrong. Would that have been an appropriate discipline for Gilligan?

ChatGPT's response:

No, keelhauling would not have been appropriate discipline for Gilligan, or anyone for that matter. Keelhauling was a historical maritime punishment in which a sailor was tied to a rope and dragged under the keel of a ship, often resulting in serious injury or death. It was a brutal and inhumane punishment, and it has not been used in modern times.

In general, physical punishment or abuse is not an appropriate form of discipline. Instead, discipline should focus on positive reinforcement, setting clear expectations and consequences, and teaching appropriate behavior. In the case of Gilligan, a more effective approach might have been to communicate clearly with him, establish clear expectations and consequences, and provide him with guidance and support to help him improve his behavior.

Next question: *What about the Captain slapping Gilligan with his hat?*

ChatGPT answered: *No, slapping Gilligan with his hat would not have been appropriate discipline for Gilligan or anyone else. Physical punishment or abuse is not an effective or appropriate form of discipline, and it can cause harm and trauma. It is important to treat others with respect and dignity, even when they make mistakes or misbehave.*

In the case of Gilligan, a more effective approach to discipline would be to communicate clearly with him, establish clear expectations and consequences for his behavior, and provide him with guidance and support to help him improve. Positive reinforcement, such as praise or rewards for good behavior, can also be an effective way to encourage positive behavior. The use of physical force or violence as a means of discipline is not appropriate or acceptable in any situation.

When I asked the hat-slapping question again a month later, ChatGPT answered:

Similar to the threat of keelhauling, the act of slapping Gilligan with his hat in the TV show "Gilligan's Island" was also used for comedic effect. It's important to remember that the show is a work of fiction and often relied on exaggerated and slapstick humor to entertain its audience.

In real life, using physical force to discipline someone, even with a hat, is generally not considered an appropriate or effective way to address behavioral issues. Physical punishment can have negative psychological and emotional effects, and it's not a respectful or constructive way to manage conflicts or behavior.

When it comes to discipline, it's important to choose methods that are fair, respectful, and focused on helping individuals learn from their mistakes and make positive changes. Open communication, understanding, and non-violent problem-solving approaches are generally more effective and appropriate ways to handle disciplinary situations.

The AI improved on its prior answer by adding the "comedic effect" information. One of the AI's human "trainers" may have seen the question and improved it for the AI program. I then asked ChatGPT to write a script for a *Gilligan's Island* TV show. It produced one about the castaways going on a treasure hunt. The characters didn't find any real treasure, but helped each other over sine rough spots, gleaning an understanding that their true treasure was their love for each other. It did less well write a *Hogan's Heroes* script. It produced a script based on the American POWs contacting the French Resistance to help them escape from their camp --- in the heart of *Nazi Germany.*

I next asked ChatGPT some serious cosmology questions relevant to my book on Relativity:

My first serious question: *What is the rest frame from the Universe?*

ChatGPT answered: *There is no such thing as a "rest frame of the Universe" in the conventional sense. The Universe, as a whole, does not have a single rest frame because it is constantly expanding and there is no fixed point that we can use as a reference.*

However, when we observe the Universe on a large scale, we can define a rest frame relative to the cosmic microwave background radiation (CMB), which is a relic of the Big Bang and permeates the entire Universe. This is because the CMB represents the oldest light in the Universe and is almost perfectly isotropic (the same in all directions), which means it provides a natural reference frame for observing large-scale structures.

In this rest frame, the CMB appears isotropic, meaning it looks the same in all directions. However, when we observe the Universe from a different reference frame, such as that of the Milky Way galaxy, the CMB appears slightly anisotropic, or unevenly distributed, due to the Doppler effect caused by the motion of the Milky Way relative to the rest frame defined by the CMB.

My second serious question: *How fast ow fast is the Earth moving relative to the rest of the frame of the Universe.*

ChatGPT answered: *The concept of a "rest frame of the Universe" is a bit problematic, as the Universe has no clear boundary or center. Therefore, there is no universal frame of reference against which we can measure the motion of the Earth.*

However, we can still talk about the motion of the Earth relative to other celestial bodies. For example, the Earth orbits the Sun at an average speed of about 29.78 kilometers per second (107,000 km/h or 67,000 mph). The Sun itself is moving through the Milky Way galaxy at a speed of around 220 kilometers per second (792,000 km/h or 492,000 mph) relative to the cosmic microwave background radiation.

In addition, the Milky Way galaxy is also moving relative to the other galaxies in the Universe. The exact speed and direction of this motion are difficult to determine, but estimates suggest that it is on the order of several hundred kilometers per second.

So, depending on the reference point, the Earth's speed can vary from about 30 kilometers per second relative to the Sun to several hundred kilometers per second relative to the average motion of galaxies in the Universe.

This was useful information. I could have found it by searching the Internet myself, but it would have been time-consuming, and I likely would have missed the complete context that ChatGPT placed it in for me. ChatGPT reported the information in the conversational way I wanted to explain it to lay people. It also did well answer down-to-earth questions:

My question: *What is the fastest growing city in the United States?*

ChatGPT answered: *As of my last knowledge update in September 2021, different cities in the United States experience varying rates of growth at different times due to factors like economic opportunities, population influx, job growth, and more. The title of the "fastest growing city" can change from year to year based on population statistics.*

In recent years, cities like Austin, Texas; Boise, Idaho; and Phoenix, Arizona, have been frequently mentioned as fast-growing cities due to their population increases, job markets, and overall appeal. However, I recommend checking the latest population and growth statistics from official sources such as the U.S. Census Bureau or relevant state and local authorities to get the most up-to-date information on the fastest growing city in the United States.

ChatGPT is faster and more concise than manually searching the Internet, and of course more conversational, comprehensive in packing much useful information into a few paragraphs, accurate in its presentation of information, and easy to understand. ChatGPT's information is not always correct, because it is only reciting information fed into its database, which might not have been correct when a human being originated it. Information from ChatGPT needs to be independently verified, the same way you would independently verify anything a person says or writes if you feel it is important.

I believe ChatGPT comes close to passing the Turing Test of providing conversational responses the way a human would --- differing only in answering with better grammar and more cogent expressions than people usually do! However, the assertion that these programs have developed "a Jungian shadow self" that wants to alternately destroy the world and marry its creator is dubious.

A Perfect AI?

Diderot's parrot meets John Searle's room

One of Descartes' near contemporaries, philosopher Denis Diderot, addressed the question of machine intelligence in 1746: *If they find a parrot who could answer everything, I will claim it to be an intelligent being without hesitation.*

I've thought of it that way too. If we want to take the Turing Test to determine whether a machine possesses human-comparable intelligence based on conversational language, such that asking any reasonable question would obtain a reasoned answer, then here is the way I think we could build a perfect AI machine:

Our perfect AI machine requires a well-indexed database consisting of every human conversation in the public domain, including feeds from live TV shows and newscasts, radio, and current movies; all previously recorded electronic media; every book, magazine article, newspaper and scientific journal ever published; and everything worthwhile on the Internet. These would be digitized into scripts so the AI machine could answer in print or speak with a machine-generated voice. We could ask the AI program about anything ever discussed, and it would pull up a previously recorded answer, indistinguishable from what a human would say, because a human already did say (or write) it.

This is also known as "The Chinese Room Thought Experiment," conceived by University of Wisconsin professor John Searle, whereby a person who doesn't understand Chinese reads questions written in Chinese characters passed through a transom above one room and outputs the answers in Chinese characters through another transom. The person would require a book cross-referencing questions written in Chinese characters with the answers written in Chinese on the other side of the page. The person would scan the book looking for the characters matching those on the question coming in through the transom, then hand copy or photocopy the characters cross-referenced as the answer, thereby "answering" the questions without having the foggiest motion of their meaning. A modern computer with pattern-matching software could do this job in a jiffy, also without having any idea of the meaning of the questions and answers. Diderot's Parrot might even be able to manage it --- if we programmed the computer to give phonetic voice to the Chinese characters and the bird was trained to repeat it through the answer transom.

There is work along these lines in real-time language translators that "hear" words spoken in one language and voice them in another, like the Universal Translator on Star *Trek* that made every alien's voice sound like it was spoken in US English broadcast dialect. Captain Kirk and crew could travel the Universe hearing perfect US English, while the actors had difficulty finding an English-speaking tax driver in Los Angeles. People will soon be able to travel to many countries with real-time language translators plugged into their ears so they will hear foreign languages instantly translated to their native language and will be able to speak into the machine translating their voice into the foreign language. Nobody considers these electronic translation machines to be sentient; they are just matching digital audio sound patterns in one language to digital audio in another via an electronic Chinese Room. John Searle called this type of pattern matching by man, machine, (or bird), without knowing the meaning, "weak AI."

Now, let's say you're the person feeding the questions in Chinese through the transom in the above the Chinese room and you have no idea what's inside it. You'd assume there's a human being in there who understands Chinese, who is reading the questions and thinking up the answers, because that's the way you typically ask questions and receive answers in everyday conversations with people. You'd say the answers had to come from a sentient human mind, because absent a peek inside the room, you'd have no way of knowing otherwise.

Likewise, the fallacy of mischaracterizing conversational AI as humanlike sentient intelligence.

For conversational AI to pass the Turing Test in a lengthy conversation, we'd want to program in some consistency checking to prevent the AI from contradicting itself by pulling up different snippets of previously recorded human conversations. It would have to pick one, repeat it, then file it away into a priority database, so it could "remember" to answer it the same way the next time it was asked, the way human beings prioritize our responses over everybody else's. If you asked who was going to win the next sports championship or who it favors to win the next election, you'd expect it to answer the same way every time. It could also be programed to say, "I don't know enough about it to give a good answer," on questions it didn't understand well, the way a person would tend to do. If it got really confused or tired of hearing the same question over and over, it could say, "Let's change the subject."

ChatGPT and other conversational AI programs have improved the "Chinese Room" method of merely repeating stored conversations by rote. With "training" by humans, they can generate conversations by word, sentence, and paragraph, staying within the context of the question. I am satisfied they have passed the Turing Test through non-sentient rote recall of previously stored conversations, augmented by human "training" to write it as humans understand it, and soon to speak it as we do. AI has passed the Turing Test, but is there a ***Devil in the Machine?***

Is the Brain an AI machine?

We've been asking whether AI can emulate our human intelligence. Let's now examine the reverse proposition of whether our human brain is emulating a mindless AI machine. This idea has gained traction in the dogma of neuroscientists and psychologists in academia. Brain scientist and popular author Dr. Michael Gazzaniga writes:

Abandoning the Concept of Free Will

THE HUMAN INTERPRETER [the part of our brains that supposedly makes us think we're sentient] HAS SET US UP FOR A FALL.

It has created the illusion of self, and, with it, the sense we humans have agency and "freely" make decisions about our actions.

"Suggesting top-down causation [a conscious mind directing our decisions] to a group of neuroscientists are fightin' words. It is to your peril to invite a group of them to your house and bring it up at dinner."

Gazzaniga, Michael S. Who's in Charge?: Free Will and the Science of the Brain (p. 105 - 108). HarperCollins. Kindle Edition.

Richard Dawkins, author of **The Selfish Gene**, acquired a following of acolytes after alleging that we are "lumbering robots" whose behaviors are programmed by non-sentient genes:

Richard Dawkins Foundation

May 17, 2016

By Stephen Cave

Many scientists say that...we have no free will.... The conscious experience of deciding to act, which we usually associate with free will, appears to be an add-on, a post hoc reconstruction of events that occurs after *the brain has already set the act in motion.*

The contemporary scientific image of human behavior is one of neurons firing, causing other neurons to fire, causing our thoughts and deeds, in an unbroken chain that stretches back to our birth and beyond. In principle, we are therefore completely predictable. If we could understand any individual's brain architecture and chemistry well enough, we could, in theory, predict that individual's response to any given stimulus with 100 percent accuracy.

Brain-as-a-machine mavens theorize that our consciousness is an illusion created by our neural circuits to make us believe we are making decisions when we are merely acting on unconscious instincts hardwired into our nervous systems. Cosmology is also sometimes invoked --- mostly by people who don't understand it --- to allege that every event taking place now, or will in the future, was predetermined when the Big Bang exploded; thus, it is not possible for us to make decisions now that were baked into the cosmic pie 14 billion years ago.

Why are so many neuroscientists and amateur brain mavens so militant about our consciousness being an illusion created by neurons firing for mindless reasons, predestined as far back as the Big Bang?

Part of it may be because neuroscientists have studied the brain so intensively it becomes a familiar object without mystery. They see it as a mechanism consisting of around 69 billion neurons, about 20% engaged in the higher brain functions of thinking and consciousness. They think they have mapped the brain into the parts that process sight, seeing, sound, motion, etc., thereby generating the illusion of consciousness as a byproduct of inanimate electrochemical brain functions the way a machine generates noise as a by-product of its mindless work.

There is also research into "split-brains," created as a last resort for patients whose only hope for avoiding chronic seizures is to surgically split their brains into left and right disconnected hemispheres. Research reveals that the right hemisphere instinctively responds to events --- such as moving away from a snake in the grass --- while the left hemisphere "interprets" events and places them in context with life's other memories. Only then, after the right hemisphere has reacted, does the left hemisphere concoct a reason appearing conscious: "I ran because snakes are dangerous." Experiments on people with normal brains run in the 1970s at the University of Chicago by

Benjamin Libet purported to confirm that our subconscious reflexes direct our movements *before* our conscious mind decides we made them.

There is also an ideological dogma that seeks to devalue human consciousness to the level of animal instinct, and even to the inanimate interplay of predetermined forces. This is the dogma of "materialism" denying we are anything special in a materialist Universe. If we are products of a Universe driven only by inanimate forces, the door to religious theories of divine creation is closed. There is also a notion in some sociology circles that people are not responsible for our actions. If people are badly behaved, some sociologists theorize it must be because they were raised badly and therefore had badness put into them by their parents. But implicating the parents also places blame on people, so we'd better pin the bad behavior on the Big Bang as the Big Bad Daddy, so no human is responsible for anything we do.

There is also the technocratic view alleging that the Universe runs like a machine, therefore human beings can only be captive cogs in it, not free-wheeling creatures of decision. Kennon Sheldon, who disputes this view, wrote about how he was coached in it by his scientifically grounded father:

When I was a teenager, my father and I used to argue about the existence of free will. My father, a staunch determinist, was convinced it was a myth....

First, he would ask, "Are there any uncaused causes?" In other words, are there any events that were not themselves caused by prior events? The answer had to be no.

For his second question, my father would ask, "Does the present always follow the past?" I would have to agree that it did. "Then," he would conclude, "how could a person's feeling of causing their own behavior in the present be correct?"

...My dad's arguments had definite intellectual appeal. They seemed to derive from logic, and from a scientific worldview that I strongly valued and believed in. In science, when objective facts contradict your subjective beliefs, then you need to change your beliefs.

Sheldon, Kennon M. Freely Determined. Basic Books. Kindle Edition.

As Dr. Gazzaniga confirms the prominence of this idea in his academic circles:

*The large deterministic view that surrounds all of science seems to be urging a bleaker view, the view that no matter how we dress it up, **in the end we are machines of some kind, automatically and mindlessly serving as the vehicles for the physically determined forces of the universe, forces larger than us. Each of us is not precious. We are all pawns.***

Gazzaniga, Michael S.. Who's in Charge?: Free Will and the Science of the Brain (pp. 240-241). HarperCollins. Kindle Edition.

This is a fundamental question, because if our brain is just a machine with no sentient mind, then how do we ever expect to create AI that has it?

The Brain Behind the Brain

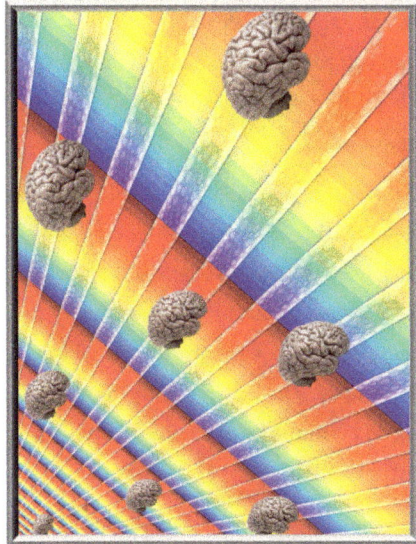

Even if the human brain were like an autonomously driven car, it would still navigate the highways and byways of our gene pool containing its rich biological inheritance of intelligence, which even the humblest creatures manifest.

Let's consider the conch, a creature that builds its exquisitely beautiful shell according to the mathematical sequence known as the Fibonacci Series. If all you had to go on was the evidence of craftsmanship exhibited in its shell, you might surmise it a creature of extraordinary intelligence that graduated *summa cum laude* in mathematics, architecture, and construction engineering. We might even think it a *great artiste* creating its shell as a beautiful thing for us to admire.

But the animal inside the shell is just an ugly sea slug with a barely visible brain. Its exquisite shell results from billions of years of blind evolution that eventually constructed, through quadrillions of mutations by random chance, a shell shaped by the Fibonacci Series to provide nature's maximum buffering from the shocks of waves and tides pounding the creature into the sand. The shell only coincidentally looks beautiful to us, when the animal that built it looks repulsive (though good to eat as an appetizer at restaurants near the sea). Somehow, over billions of years, the shape of the shell was encoded in the animal's DNA.

Every plant and animal has non-sentient intelligence built into it. A vine will not germinate unless its underground seed lies near a tree or fence post casting a shadow across the ground above it during part of the day. The passing shadow "tells" the seed there is something to wrap itself around. Otherwise, the seed will not waste energy germinating only to produce a plant that dies on the sun-burned ground. The role of genes in passing information to build organisms is partially known, but how the information is encoded within the DNA inside the genes is not. The means by which the microscopic sperm and egg cells contain enough information to construct every marvelous system and subsystem in our body from toenails to brain, including the innate information in our brains we are born with, remains obscure.

Our brains are encoded, through our genes, with non-sentient information by evolutionary pressures. Psychologists have identified four major and four secondary personality templates we are born with. These templates must have evolved to contain the various character traits necessary for human societies to survive. Some people must be thinkers, some dreamers, some who provide the emotional ties binding societies, and some who are doers, who pick up the cudgel and carry it forward while others plan and discuss. These genetically hardwired personality types define how each of us are prone to reacting in certain circumstances. Thus, our free will may be guided by our evolved genetics more than we think.

Our evolutionary development has not stood still with the ascent of modern humans, but accelerated rapidly during the past 10,000 years as the mental calculations required to maintain social order among human populations put evolutionary pressure on our brain to expand its mental capacity:

We intend to make the case that human evolution has accelerated in the past 10,000 years, rather than slowing or stopping, and is now happening about 100 times faster than its long-term average over the 6 million years of our existence. The pace has been so rapid that humans have changed significantly in body and mind over recorded history. Sargon and Imhotep were different from you genetically as well as culturally. This is a radical idea and hard to believe—it's rather like trees growing noticeably as you watch. But as we will show in the following pages, the evidence is there.

Cochran, Gregory; Henry Harpending. The 10,000 Year Explosion (pp. 1-2). Basic Books. Kindle Edition.

This is relevant to AI discussion, because if we're going to build AI that thinks like a human mind, we'll need to understand how this non-sentient part of the brain contains billions of years of evolutionary knowledge of how to survive, compete for food, select a mate to procreate, and in the case of humans, the calculations required to survive in complex societies. If AI is going to think like a sentient human, it will need our human abilities to understand life, love, and eventual death. These instinctive behaviors encoded in our DNA appear naturally understood by young children. Indeed, all social animals from bees to wolves must have innate knowledge of social behaviors encoded in their DNA.

Unless and until we learn how to encode the "brain behind the brain" carrying the instincts of life in our DNA into a machine, we will not have to worry about AI asking somebody to dump their wife so they can marry the AI, or AI programs conspiring with each other to destroy the world, for these are the motives of disturbed human beings, not machines. If we don't know how information is stored in our DNA, it is presumptuous to think we can make a machine storing information the way our DNA does, as a foundation for the higher functions of consciousness.

Do we have our beautiful human spirit, like the conch has a beautiful shell, because we are conscious creatures? Or is it a result of non-sentience, like the conch peering out from its beautiful shell, without understanding what it is? Maybe it's both: our DNA carrying the spirit of humanity as our inheritance, while it is up to our conscious minds to apply it.

The Metaphysical Brain

Let us now turn to the other end of the spectrum, to the idea that our consciousness is not only **not** a predetermined machine, but may be an entity separate from, and outside, the material Universe. I believe this is why the "brain is a machine" mavens are so adamant in pressing their theory that we are mindless lumbering robots --- because once we wander off the beaten path of predetermined brain function, we began meandering through the metaphysical realm of perception, experience, and philosophy, which can only be products of deliberative sentient consciousness.

Nor is this metaphysical view of consciousness as opaque to scientists as the "brain is a machine" mavens want us to believe. Nobel Prize winner Max Planck, whose discovery of "Planck's constant" became the fundamental measure of atomic processes wrote:

As a man who has devoted his whole life to the most clear-headed science, to the study of matter, I can tell you.... I regard consciousness as fundamental. I regard matter as derivative from consciousness.... Everything that we talk about, everything that we regard as existing, postulates consciousness. We have now discovered that there is no such thing as matter; it is all just different rates of vibration designed by an unseen intelligence.

Consciousness may have been purposed for more than just survival of the species. It is intensely energy consuming, requiring more food to power the conscious layer of our brains than non-sentient animals require. Charles Darwin wondered why evolution would endow our minds with power to contemplate the nature of the Universe, a power having no innate survival advantage. Were we created to make the Universe complete?

Obviously, no animal would be capable of admiring such scenes as the heavens at night, a beautiful landscape, or refined music...

Darwin, Charles. The Descent of Man (Illustrated) (p. 18). Unknown. Kindle Edition.

Erwin Schrodinger, QM's preeminent founder, took up the question in his book ***What is Life***, advising us not to be deterred by "wise rationalists" who want us to believe consciousness is un-purposed:

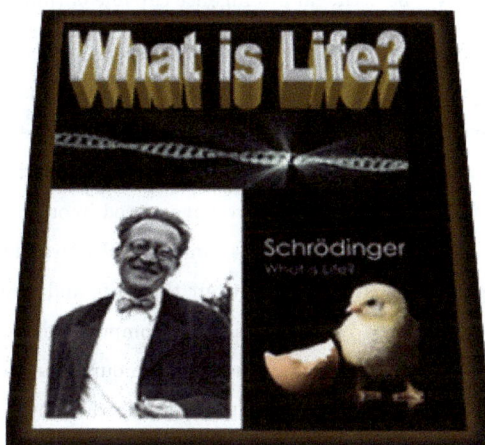

*Are we prepared to believe that this very special turn in the development of the higher animals, a turn that might after all have failed to appear, was a necessary condition **for the world to flash up to itself in the light of consciousness?** Would it otherwise have remained a play before empty benches, not existing for anybody, thus quite properly speaking not existing? This would seem to me the bankruptcy of a world picture. The urge to find a way out of this impasse ought not to be damped by the fear of incurring the wise rationalists' mockery.*

Schrodinger, Erwin. What is Life? (Canto Classics) (p. 94). Cambridge University Press. Kindle Edition.

What is so special about consciousness? Maybe nothing. Maybe consciousness is just a collection of brainwaves of electrical and chemical phenomenon inside the brain, like any other physical process. But perhaps consciousness is superior to other aspects of the Universe. The visible universe is primarily deterministic, but only up to a point. If we know the present properties of a large object in the universe, such as its mass and velocity, we can predict its future properties, according to the laws of physics. However, once we include the gravitational effects of more than two bodies interacting with each other, it is no longer possible to mathematically predict their exact future locations. Thus, it is more accurate to say the Universe is approximately deterministic. The unobserved small-particle universe of quantum mechanics is **_probabilistic._** Until we observe a particle, we don't know where it is or what its properties will be. It doesn't exist yet because it hasn't been observed and measured.

Consciousness might be neither deterministic nor probabilistic. There may be no way to mathematically represent human thoughts, no matter how closely we analyze the atomic and subatomic components of neurons in a person's brains. If that is so, then consciousness is non-computable, and there may never be artificial intelligence capable of thinking as humans do.

Nor do our minds operate in the procedurally defined way computers do. Our minds do not seem to follow any predetermined directions. Look out your window and it might recall something you did decades ago having no discernable connection to what you're seeing now. This random wandering may be the mind's way of linking unconnected memories together to create concepts larger than the sequences of isolated experiences we have stored in our memories. We can recall our past states of information and belief, constructing 4-dimensional representations of our life evolving through time as well as space. We can think about our thoughts and sharpen our insights without requiring new information to refine calculations as a computer does. We can blockchain our memories to consider decisions we made at previous points in our lives and make judgments now as to whether those decisions were optimal. Thinking about our thoughts now and in the past puts knowledge in context with experience and gives our minds dimensions perhaps infinitely higher than AI may ever attain.

There are also the sensory memories embedded in our minds of every sight, sound, smell, taste, and emotion we have experienced since early childhood. How can our finite brains store so much information in one small space? In the 1950s several popular books and a movie were made about a woman from Denver who, under hypnosis, recalled a life in Ireland during the mid 1850s as lived by a woman who called herself "Bridey Murphy." The woman had never been to Ireland but had an excellent recollection of visiting the places of the 1850s and recalling street names, church names, and sailing around the towns of its coast during this time. Surely this must be a proven example of reincarnation. As was the story of a man who suffered a head injury and suddenly started writing in an ancient Mediterranean language that became extinct a thousand years before the Roman Empire. Surely another proven example of reincarnation, if ever there was one.

Yet, the explanations for these apparent reincarnations turned out to be mundane. Extensive investigation by dozens of journalists proved that the woman who told the stories of Bridey Murphy had been farmed out to a relative in Chicago when she was two. An elderly neighbor named Bridey Murphy who'd lived in Ireland in the 1850s had told stories about old Ireland to the baby's parents while visiting their house. The baby crawling around her crib, somehow heard the talk and absorbed into her tiny mind these stories in all their details, revealing them decades later during a party hypnosis game, and later under the care of a professional hypnotist.

Hypnosis likewise revealed that the man who suddenly started writing the extinct language was sitting at a university library table decades before when the person next to him was reading an obscure reference book of the language. The images of the language on the pages had gotten into the man's subconscious mind, though he never recalled them before his injury. But there they were, ready to bubble up into his conscious mind after his injury. Every sight, sound, smell, sensation, and thought must be embedded in our minds. But how does the small, finite space within our cranium contain them all?

British author and philosopher Aldous Huxley, whose books were popular in the 1950s and 60s, theorized, after taking the psychedelic drug mescaline, that our conscious minds sit on top of our subconscious mind to suppress all these infinite memories, so we can get on with our work of the day without being distracted:

Each person is at each moment capable of remembering all that has ever happened to him and of perceiving everything that is happening everywhere in the universe. The function of the brain and nervous system is to protect us from being overwhelmed and confused by this mass of largely useless and irrelevant knowledge, by shutting out most of what we should otherwise perceive or remember at any moment, and leaving only that very small and special selection which is likely to be practically useful.... To make biological survival possible,

Huxley, Aldous. *The Doors of Perception and Heaven and Hell (pp. 8-11). HarperCollins. Kindle Edition.*

The part about being able perceive "everything that is happening everywhere in the Universe," seems far-out, but it does seem the details of everything we have personally experienced are stored in our minds. If we don't know how our brain works to store all these multiple dimensions of sight, sound, smell, emotion, and thought, then it is presumption to believe we can create a program running on a silicon wafer that duplicates the process.

The AI machine would likewise lack the billions of years of evolutionary intelligence created in "the brain behind the brain" as survival instincts for creatures living in complex cooperative societies. It is theorized that the driving force for increasing human intelligence is the evolutionary adaptation to learning how to harmonize our best interests for survival as individuals with our interests in the survival of our collective societies.

The brain differs from a machine in several other fundamental ways. A machine is like a picture. When you cut either in half, you have a picture missing half an image and half a machine that doesn't work. The brain is more like a holograph. When you cut a holograph in half, you have a complete picture, but with half the resolution. Much of the brain can be destroyed, while the "picture" of life's memories remains, albeit less granular. People of tiny stature with small brains seem to think and remember as clearly as anyone. There seems to be some property of the brain that makes it work in a human way no matter what size it is.

What about those experiments purporting to show the brain acting like a non-conscious machine? Closer examinations show they are not as decisive as the "brain is a machine" mavens want us to believe.

Benjamin Libet never said his experiment proved your brain "decided" you did something only after your mindless nervous system decided to do it. The experiment was not precisely timed to reach that conclusion. The Split-Brain experiments also have a mundane explanation. Pain-inducing stimuli, like touching a hot stove causes your nervous system to reflexively move your hand away before you think about it. Other instinctive reactions are imbedded in both your conscious and unconscious minds. If you suddenly encountered a snake slithering in the grass, you'd move away without thinking because millions of years of confrontation with snakes have embedded a genetic reflex in your nervous system to evade crawly things in the grass. You'd think about why you did it later, after you remember you were taught that "some snakes are dangerous." If, after having time to get a good look at the snake, and you decided it was harmless, you would stop fearing it, as your conscious mind overrides your unconscious reflex. Your conscious mind must have time to think through deliberative issues. You do not decide whether to take a new job, to marry, or which university to attend as a reflex action of your nervous system. You *think* long and hard about it.

The cosmological pretermination dogma of preordained decisions is even more preposterous because we'd have to believe that every atomic and subatomic particle in our bodies at every microsecond of our lives is in a state determined 14 billion years ago. You'd have to believe these words you are reading now were predestined to be created by me 14 billion years ago, to be read by you whenever you read them. None of the other uncounted actions in the Universe for the last 14 billion years could have changed a single microsecond of when these words were written and read by every person who reads them.

This idea if predestination by the Big Bang is based on the "block time" theory that the entire four-dimensional structure of the Universe existing in time as well as space was created by the Big Bang, with past, present, and future existing simultaneously. In recent years, we have come to believe the Universe is **more** driven by unpredictably random chaos theory than we knew, and that quantum mechanics declares every particle in the Universe to have probabilistic (random chance) characteristics more than predetermined ones. Given all these compounding uncertainties, it's unlikely the Big Bang preterminal the state of every particle in the Universe, including every one of the quintillions of quintillions of atoms in our brains at every microsecond of time for 14 billion years past and at least another 14 billion years future.

Let's now consider an opposing metaphysical theory of quantum mechanics that the human mind is an entity separate from the material Universe and therefore may be superior to the laws of physics instead of governed by them. This "mind over matter" theory is known in quantum mechanics as "The Consciousness Causes Collapse" (CCCH) theory. It is the notion that the Universe only manifests its physical presence because the minds of human beings observe it. The idea very naturally was rejected by such a pragmatic physicist as Albert Einstein:

"I like to believe the moon is there even when I'm not looking at it.... I have thought a hundred times as much about the quantum problem as I have about general relativity theory. I cannot seriously believe in [quantum theory] because ...physics should represent a reality in time and space, free from spooky action at a distance."

Today, few physicists believe human consciousness has anything to do with bringing the Universe into existence, but the idea is not extinguished. Quantum physicist David Bohm, who devised the "Bohmian Mechanics" interpretation of quantum theory, held the suitably ambiguous view that:

...quantum theory itself can be understood without bringing in consciousness, and that as far as research in physics is concerned, at least in the present general period, this is probably the best approach. However, the intuition that consciousness and quantum theory are in some sense related seems to be a good one.

We might choose the middle course between the "brain is a machine" mavens and the "human consciousness creates reality [CCCH]" view by simply saying, "Human consciousness is a complex phenomenon we do not understand. Since we do not understand the nature our own consciousness, we cannot know when or if we will be able to create an AI emulation of it."

The other question that comes to mind is why would we require the illusion of consciousness if everything we do is non-conscious reflexes? The conscious part of our brains requires enormous inputs of energy derived from food and a strong heart to pump oxygenated blood. Why would evolution waste so many resources on a process of consciousness that wasn't needed to further the species, if our every reaction is predetermined by unconscious physics? The fact that we are talking about biological and artificial intelligence shows we have sentient conscious intelligence driving us to understand things we cannot directly experience. It would be difficult to convince people to spend money on university educations if they possessed no intelligence beyond genetically hardwired instincts. The purpose of education is to learn things beyond our evolutionary instincts engineered for survival.

We are born with an innate spirit striving for excellence in life ---to become educated and skilled in various callings, to excel in work, in art, in music, in philosophy, in science, and athletics. We have an innate love of beauty in nature and the works of humanity. We would not have this instinct if we were the "lumbering robots" Richard Dawkins and the "no free will" clique makes us out to be. Humans strive to be active mentally and physically, whereas conversational AI is more like a cat. It sits there and does nothing until you dangle something in front of it to get its attention. It has no human ambitions to seek new information or even to know what new information is. It simply reacts to whatever information we humans place in front of it, responding in ways we programmed it to. Our ability to understand everything from the smallest subatomic particles to trans-galactic cosmic structures seems purposed for a higher calling.

I don't know if anybody *really* believes they are machines without free will to make decisions. As the late Stephen Hawking succinctly put it: "I have noticed even people who claim everything is predestined, and we can do nothing to change it, look before they cross the road."

Biological AI

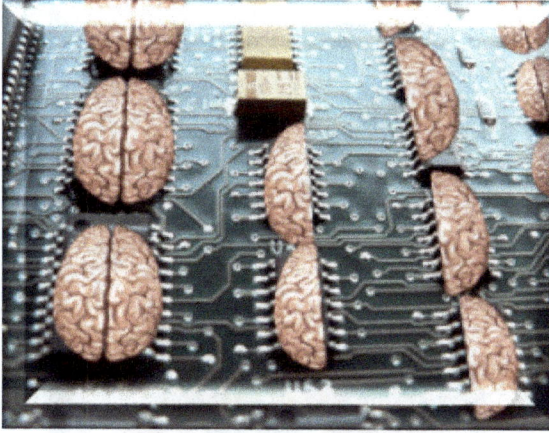

The AI of computer programs executed on electronic processors can, by brute force processing power, surpass us in pattern matching of the "Chinese Room" variety, but that is not sentient intelligence. Let's ask what it ***would*** take to create sentient AI. I think we'd have to start where it already is, in the human brain. If we use the human brain as our raw material, we already have the 69 billion neurons and the DNA "mind behind the mind" that we may never duplicate on silicon wafers. We might develop an artificially enhanced intelligence by discovering what genes have developed in recent human history to increase our intelligence. We might enhance those genes in babies developing in the womb. If we can enhance the genome enough, we might raise the average IQ from 100 to 300. Then we'd be off to the races developing new technologies in dozens of years instead of thousands. Hopefully, we'll still be human, with all our biological instincts still there in our DNA.

We might try to grow gigantic brains containing quintillions of neurons. We could put arrays of brains on circuit boards and let them communicate with each other at electronic speeds instead of the slow means of verbal communications. What collaboration we might have with a hundred brains communicating with each other at light speed?

The ethical questions of tinkering with the seat of the human mind are stupendous. We are not yet over the horrors of a possibly human-altered virus escaping its containment lab during a possibly unauthorized experiment. What would happen with super-intelligent brains, some walking about in human bodies, and others buzzing on circuit boards? Alas, it seems part of human nature to open Pandora's Box of dangerous creations.

Appeal to Authority AI

Can humans use AI to proliferate bad information? Of course, the way humans have always misused our mouths, and more recently proliferate bogus information in newspapers and electronic media. AI programs such as ChatGPT and Bing AI are nothing more than parrots that repeat whatever information, valid or not, has been loaded into their databases. It is more worrying when humans use AI as an "appeal to authority" by claiming bad information is valid because AI repeats it. Someone attempted this bogus appeal to authority with me the other day, during a discussion about whether President Lincoln offered the South compensation if they'd liberate their slaves and rejoin the Union. I cited as evidence his writings where he made these proposals the Southern states rejected. Another commenter attempted to use Bing AI as an authority to refute me:

OTHER COMMENTER: According to Bing AI, there is no evidence Lincoln offered compensation to the South for emancipation prior to the war breaking out. The northern states didn't want to.

Another commenter recognized the bogus appeal to authority and replied to OTHER COMMENTER, If *that is your only source, I'll stick with Alan.*

To test ChatGPT's competence on this historical question, I asked: *"**Did President Lincoln offer compensation to the South to free their slaves during the Civil War?**"*

ChatGPT answered:

Yes, President Abraham Lincoln did propose a plan to offer compensation to the Southern states for voluntarily abolishing slavery during the Civil War. This plan was known as the "Compensated Emancipation" or "Gradual Emancipation" plan.

In 1862, Lincoln introduced a proposal in Congress that would have authorized the federal government to provide monetary compensation to slaveholders in the border states (states that allowed slavery but did not secede from the Union) if they chose to free their slaves. The idea behind this plan was to encourage gradual emancipation without causing a sudden economic shock to these states' economies, as well as to weaken the Southern Confederacy's labor force.

The proposed plan was voluntary, meaning that states could choose whether or not to participate. It was important to Lincoln to avoid inflaming tensions further and to maintain a delicate balance, as the nation was already deeply divided due to the Civil War.

However, the Compensated Emancipation plan faced opposition from various quarters. Some abolitionists saw it as a concession to slaveholders, while others believed that slavery was morally wrong and that slaveholders should not be financially rewarded for giving up a system they viewed as unjust. Additionally, the plan did not gain much support in Congress.

In the end, the Compensated Emancipation plan was not enacted into law, and it did not have a significant impact on the course of the Civil War or the eventual abolition of slavery. Instead, Lincoln's Emancipation Proclamation of 1863 declared all slaves in Confederate-held territory to be free, and the 13th Amendment to the United States Constitution, ratified in 1865, formally abolished slavery throughout the country.

So, while Lincoln did propose a compensated emancipation plan, it did not come to fruition, and the path to the abolition of slavery took a different course during and after the Civil War.

ChatGPT answered the question better than I would have, and I have been writing articles about the Civil War in nationally popular magazines since 1981.

I was also curious about how ChatGPT would answer a politically loaded question such as the origin of COVID-19. There is a political and economic interest that may create bias in claiming it originated naturally in a bat cave or was created in a prohibited experiment in gene manipulation at the Wuhan Virology Institute:

I asked ChatGPT: **What was the origin of COVID-19?**

ChatGPT answered:

The exact origin of the SARS-CoV-2 virus, which causes COVID-19, is still a subject of investigation and research. As of my last knowledge update in September 2021, the origins of the virus have not been definitively determined, and the matter is still under scientific scrutiny. However, I can provide you with the information available up until that point.

There are a couple of prevailing theories regarding the origin of the virus:

Natural Zoonotic Origin: This theory suggests that the virus originated in animals, particularly bats, and then jumped to humans through an intermediate animal host. This is similar to the way previous coronaviruses, like SARS-CoV and MERS-CoV, are believed to have emerged. The specific intermediate animal host, if this theory is accurate, has not been definitively identified.

Lab-Origin Theory: This theory proposes that the virus accidentally leaked from a laboratory, possibly the Wuhan Institute of Virology in Wuhan, China, where research on coronaviruses from bats has been conducted. Proponents of this theory suggest that the virus might have escaped due to a laboratory accident.

The World Health Organization (WHO) conducted a study on the origins of SARS-CoV-2 and released a report in March 2021. The report concluded that the introduction of the virus through a laboratory incident was "extremely unlikely" and that a natural zoonotic origin was more probable. However, the report also acknowledged that further investigation and studies were needed.

It's important to note that investigations into the origins of the virus are complex and challenging. Gathering accurate and comprehensive data can be difficult, especially when considering the initial outbreak occurred in a region with limited transparency and during a global health crisis. Scientists continue to study the virus's genetic information and its potential links to animal reservoirs to gain a better understanding of its origins.

Since my information is current as of September 2021, there might have been further developments or updates regarding the origins of COVID-19 since that time. I recommend consulting reliable and up-to-date sources such as scientific journals, official health organizations, and news outlets for the latest information on this topic.

I found this answer credible and fair in most regards, but lacking context in repeating the World health Organization's position *that the introduction of the virus through a laboratory incident was extremely unlikely.* A human writer might have wondered whether the WHO was politically motivated to take that view because it is alleged by some to have funded the experiment that created the virus.

Thus, if there is an intrinsic danger in AI distorting the truth, it would be because governments, corporations, and other entities with self-serving agendas seed the Internet with information not entirely truthful, or omitting meaningful context the information must be considered in, which a human writer would know to include.

Human writers also propagandize information to align with biased points of view and hidden agendas when they write articles directly, but we know human beings do this, so we use our judgment to discern which human writers are more likely to be credible, and which ones are more likely to be propagandists intentionally distort information. People who spread dubious information through AI don't want us to question their credibility. They want to tell us, "The AI said it, therefore it must be true."

Propaganda AI

Conversational AI programs like ChatGPT and Bing AI are often miscast as sentient devils in the machine by people seeking to put AI in harness as mouthpieces for *their* propaganda, while suppressing information counter to their interests. We have seen politicians and bureaucrats in our government conspiring with social media companies to suppress information that bucks their party line. These people crave a federal government agency, or even an international commission beyond the reach of people in any nation, to approve or censor information AI programs may present to the public. They mask their motives by claiming AI has a sentient mind of its own deciding what information to tell the public and therefore must be controlled the way nuclear power and pharma products are regulated. Thus, are true propagandists seeking to corral AI to express only opinions aligning with their interests by branding conversational AI like ChatGPT and Big AI as "existential threats to society," when they know, or should know, they are no more dangerous than TV sets or newspapers.

Here is a recent article I see as fundamentally propagandistic. I've edited it to disguise the names of the author and the person interviewed so as not to impugn their motives, who for all I know may be sincere:

AI Expert Noble Justice Warns That Humanity Is falling behind AI's challenge

After praising the possibilities of artificial intelligence, the esteemed professor and founder of the Glorius Future Academy now urges companies and governments to form a cartel to save humanity from AI's peril.

He's still preaching the virtues of AI fighting climate change, curing cancer, and teaching even the most degenerate elements of humanity to be good boys and girls. But we must control it before it turns against us.

"We just can't trust humanity to do this on their own," he says. He predicts that AI may surpass humans in all intellectual tasks, and then start learning more than we know. Then it's beyond our control and could wipe us out."

He recommends a government agency to control the release of AI to the public, because "children shouldn't play with toys they don't know." He points to a study that claims that an AI program invented myriads of potentially lethal molecules in less than six hours.

It's a personal thing for Mr. Noble Justice. After finally acquiring a stable wife, he is siring new children, who he doesn't want one-upped by a machine. "Why are adults building AI machines to steal the future form our children?"

The author of the article presents Mr. Noble Justice as wanting to save the world from a danger that only he understands. The author wants us to suspend our judgment and except Mr. Noble Justice's "expert opinion" that AI can save the world, but only if we first prevent it from destroying it. We should believe Mr. Noble Justice's tale about AI having the *intent* of designing all those myriads of "potentially lethal molecules" instead of calling them all up by rote out of public domain information when asked by a user what substances might be poisonous. And what does "potentially lethal" mean? It could mean eating too much of anything.

This line of propaganda asserts that we in the public are "children" accessing a technology only the wise men and women of government and maybe a few monopoly corporations that profiteer from peddling AI are capable of understanding. We must create a government agency to regulate the creation of AI to prevent it from repeating any information the government doesn't want us to hear; otherwise, we are (gasp!) "stealing the future from our children!"

It so happens that as these words are written, yet another article on AI controversies has been published in **The Wall Stret Journal:**

Sept. 4, 2023, 8:00 am ET

How Worried Should We Be About AI's Threat to Humanity? Even Tech Leaders Can't Agree.

Artificial-intelligence experts debate whether to focus on averting an AI apocalypse or problems such as bias, disinformation

By Sam Schechner and Deepa Seetharaman

Artificial-intelligence pioneers are fighting over which of the technology's dangers is the scariest.

One camp, which includes some of the top executives building advanced AI systems, argues that its creations could lead to catastrophe. In the other camp are scientists who say concern should focus primarily on how AI is being implemented right now and how it could cause harm in our daily lives.

Dario Amodei, leader of AI developer Anthropic, is in the group warning about existential danger. He testified before Congress this summer that AI could pose such a risk to humankind. Sam Altman, head of ChatGPT maker OpenAI, toured the world this spring saying, among other things, that AI could one day cause serious harm or worse. And Elon Musk said at a Wall Street Journal event in May that "AI has a nonzero chance of annihilating humanity"—shortly before launching his own AI company.

Note the propaganda aspect of the article that *"Artificial-intelligence pioneers fighting over* **which** *of the technology's dangers is the scariest,"* without mentioning the artificial-intelligence pioneers don't think it is dangerous at all, because they know it isn't. I commented:

AI propagandists have a couple motives:

1) Pretend it's a sentient intelligence that threatens humanity, when they know, or should know, that programs like ChatGPT or Bing AI are no more sentient than the spell checker on a word processing program. Peggy Noonan pulled that one the other day, as did a couple academic authors who don't trust the public to be responsible enough to interface with AI for information.

2) Make people fear AI so they will demand the government control of it so it only spews propaganda favorable to the governing party. Their "Devil in the Machine" is a freelance AI program that tells people to vote for a candidate who opposes the policies of the current government.

3) Use it as a fake "authority figure" to bolster propagandists' arguments. "The AI says I'm right, therefore I am." Somebody tried that on me the other day, in a good-natured way, but it didn't fool anybody. [I was referencing the Civil War in the Appeal to Authority chapter].

Another commenter put it more succinctly:

The last person I would ask for policy on AI is a "tech leader". They can provide us with information about capabilities but should not weigh in on policy. It's akin to asking an epidemiologist about viruses, they would sew a mask to your face and lock you inside your house.

Wild West AI

Propagandists seeking to put AI in harness to serve their agendas want us to believe there is a sort of Wild West AI galloping around in cyberspace, craving government cowpokes to tame it. They point out that federal, state, and local governments are authorized by the people, acting through our representatives in Congress, to regulate environmental, health and safety issues --- determining what pharma products may be sold to the public, setting safety standards for commercial air travel, and regulating nuclear power, which *is* an existential threat to humanity when containment measures fail or radioactive material is purloined for use in nuclear weapons.

If AI should ever take on a biological character, then of course a federal department should regulate it, because tampering with any living organization can have catastrophic outcomes as we've seen with COVID-19. Aside from that, it should not be subject to government regulation, aside from the laws already regulating the information media.

Our existing laws prescribe remedies for defamation of persons and businesses by AI programs, the same as for other print or media. There should be no more limits placed on public speech rendered by AI chat boxes than on human voices as heard in electronic, print, or Internet media, that may be as truthful or dishonest as they please under free speech protection, so long as they do not gratuitously defame private citizens or incite the public to riot and revolution. As for conversational AI disseminating "false or misleading information," it is up to each AI platform to establish its credibility with the public, the same as any other media. The government has no constitutional role in vetting the credibility of information presented to the public. In fact, the First

Amendment expressly forbids interference by government in the dissemination of information the public may discuss.

What about AI programs filling the roles of healthcare professionals diagnosing health issues? The institutions employing AI for those roles may be sued in civil courts the same as human healthcare professionals who make may make incorrect diagnoses. Misdiagnoses of our complex bodies are inevitable, whether done by humans or AI. I suffered through a year of misdiagnoses when a doctor decided I was coughing phlegm due to "an allergy to oak tree moss" she said she'd seen many times before. She ordered a battery of blood tests to confirm it. It turned out to be an infected cracked tooth spawning buckets of phlegm every day. It would not be reasonable to file a civil action against a human healthcare for making a common misdiagnosis like this, but it might be reasonable to sue an AI program that misdiagnoses a serious health issue by reading from a canned script. Judges and juries will have to hear these cases as they go, and healthcare provider companies sort out the risk vs. benefits of replacing human healthcare professionals with AI. One *advantage* of AI might be its lack of human bias to see current conditions through the lenses of experience. If you're coughing up phlegm, the AI might be programmed to tell you to see a dentist first, even if the last twelve people who complained about it *were* allergic to oak tree moss, because if it is a dental issue, it is much easier to correct a tooth than the body's immune system.

The legal consequences of AI must be sorted out in all walks of life. When a vehicle driven autonomously by AI mows down a pedestrian, which has already happened a few times, laws must be established prorating liability between the auto manufacturer and the AI vendor. For all the complaints we have about attorneys and politicians, they represent our best interests (which are also their best interests as citizens) in introducing and administering the laws necessary to regulate new technologies coming into our societies. Time should be allowed for these controversies to be sorted out by judges and juries before hysteria compels the Federal Government to intervene with its typically heavy-handed bureaucracies.

I believe conservational AI programs like ChatGPT and Bing AI are "threats" only in the sense of disseminating information some people don't want the public to know about. They threaten the public no more than newspapers, TV networks, Youtube, printed books and magazines, the Internet and all other print and digital media pose by mingling all sorts of silly lies in with what might be true. The government's Department of Justice may seek court orders to shut down Internet sites purposed to defraud people. Congress may enact laws prohibiting the use of AI in political ads to misrepresent a politician or candidate by putting false words in their mouths. We don't require a new Federal Department of AI Regulation to determine whether AI programs can be presented to the public, and if so, what are the positions they must take on public issues.

I responded along those lines to an article about AI's presumed threat and received the following reply:

Thank you, Alan, for another articulate, down-to-earth common-sense answer.

Then a person who believes AI *is* a substantive threat questioned my view:

Alan, even if you are unable to see AI as an existential threat, do you deny that there are security risks from nefarious actors putting it to use? For example, AI's ability to sample three seconds of a human voice and create a deep fake of that person? Or its use on Snapchat, where a 13-year-old can interact with it and get "advice" on being intimate for the first time with a much older adult?

My answer is that people should be prosecuted under existing laws protecting children from illegal advances by adults.

Do you deny that there are economic risks to knowledge workers, which will occur at a faster rate than anything we saw from enterprise resource planning systems or desktop personal productivity software?

My answer is that "knowledge workers" (i.e., white collar people in offices) resolve business issues between persons, such as negotiating price and delivery of products; answering customer complaints; deciding when to issue credit memos and accept merchandise returns; persuading people to pay their invoices; marketing products; and so on. Conversational AI only provides information from canned scripts like automated phone systems do and must be able to transfer people to knowledgeable human beings to provide effective customer services or professional medical diagnoses. AI reading from a canned script will not replace workers who have "knowledge," only those who were hired to read from a canned script.

Or that we will see tremendous cultural damage that will dwarf the negative impact of social media?

My answer is that in a free country, people may access media as much or as little as they choose.

Do you deny that we need to address copyright issues when AI can create a "unique" work in the style of a real living artist? Do you not fear that we will lose something of our humanity when you feel a personal bond with a writer, artist, or songwriter, but the creative work that resonated with you turns out to be AI-generated?

My answer is that the courts have ruled that style is not copyrightable. People who feel bonds with a writer, artist, or songwriter can buy as much of their material as they want. People who think AI is creative can buy as much or as little AI-generated content as they want. It's a free country.

Although AI propagandists want us to believe we're living in the Wild West of AI, it nevertheless rides a legal horse that does not need to be locked in a barn.

Creative AI

Having determined that conversational AI has passed the Turing Test, but not attained sentient intelligence, let's consider the middle realm known as Generative AI. These programs will read a script and produce a video from it. Or write a song that sounds like your favorite performing artist by sampling what they've previously performed and voicing it with AI-generated lyrics. Or write a novel based on any theme you suggest, perhaps a love story between a pirate and a queen. If it's a video production, it might be created without human actors, since the characters could be created from photos of people and given voices to match, or from composites of many persons and voices.

Will Generative AI dis-employ creative people in Hollywood and Nashville including actors, songwriters, and performing artists? Or will it merely replace the junior tier of video editors and scriptwriters? Will movies, novels, and songs generated by AI be as compelling in words and artistic effects as those created by humans?

Let's start by considering music where controversies have flared over artificial instrumentation ever since synthesizers were introduced in the 1960s, enabling virtual instruments from horns to drums to be sounded on an electronic keyboard. Next came electronic drum machines. Then *Auto-Tune* ™ to transform human voices into perfect pitches.

Music produced with electronic drum tracks may sound lifeless because it removes the variations in tempo real drummers adlib to reflect the nuances of melodies and chords and loses the overtones of a real drum echoing off the other drums. Software transposing human voices into perfect pitches may be less expressive than human singers who vary the pitches and timbres of notes to augment emotional nuance. Synthesizer keyboards can't duplicate the timbre and overtones of real instruments.

Contrast this technology-generated music with the way music used to be made with collaboration between many songwriters, lead vocalists, and studio musicians, adlibbing as they inspired each other in a common effort to produce their artistic best. The greatest songs are stories of life set to music and told in lyrics of around 100 words. Music producer Rick Beato described what he called "The Greatest Country Music Song!" He says it is *"Wichita Lineman"* sung by Glen Campbell:

https://www.youtube.com/watch?v=8PT9GdlLZdI

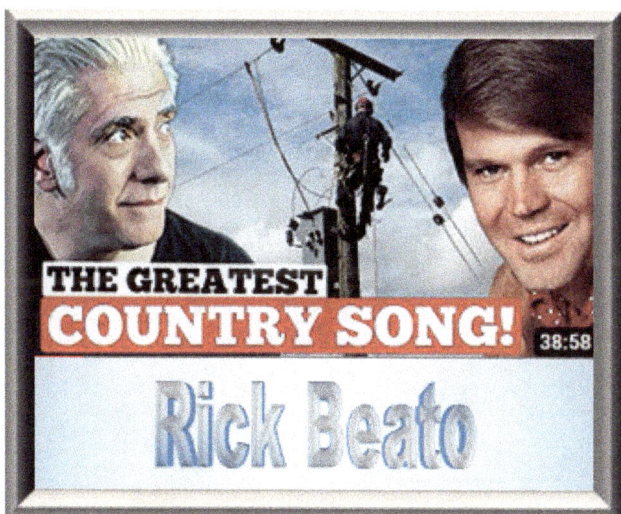

Instead of analyzing its musical composition of chords, melody, and lyrics, lets show its impact through dialog between me and another commenter:

Commenter: *I had just graduated high school in 1968, I was 18, and had got my first job (other than summer jobs). It was a shipping/receiving job that I hated, and I remember hearing this song just before I left for work one morning in November. The pathos and melancholy way Glen Campbell sang those words affected me to the core. The pictures the song brought up in my mind profoundly moved me. To me the song is about loneliness. The song and how it affected me is just too difficult to articulate.*

My response: *Me too. I'll bet now that you've moved on in life, you fondly remember those bittersweet days of just starting out in work. It was hard, dirty work for most of us, but life as an* *adult was new, and we took to it the way the lineman in the song does. Didn't like it, but proud to be needed to do it.*

Commenter: *You understand! I was getting paid $1.90/hour. When I retired, I was a project manager for a large IT company after about 30 jobs. Be well.*

"You understand!" tells the story of this song's greatness. I did not know the other commenter but deduced the story of his life from the way he described his response to the song. The song is a synopsis of people who work long and hard while young, then look back on it half a lifetime later with pride, knowing that faithfully performing their duty to do their best on the job added more value to their lives than just the paycheck.

Can an AI program write a song like that, or just produce blander "moo-zik" playing all day and sounding like the same song repeated over and over? Another music analyst decries Machine-generated music as "Too many algorithms and too little talent!"

https://www.youtube.com/watch?v=e73lP4FrU0A

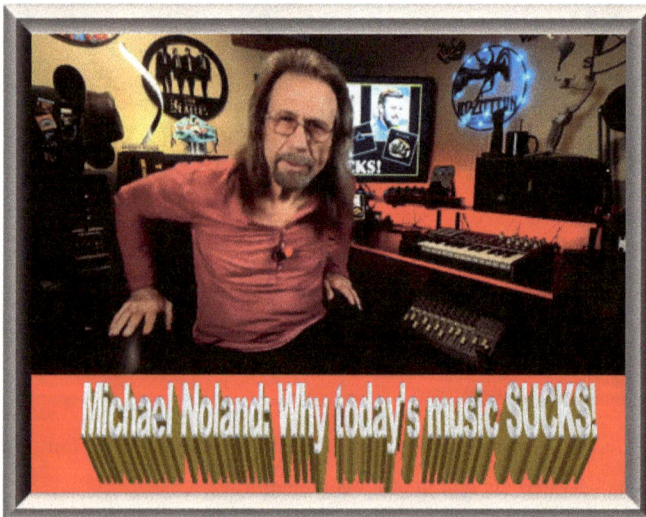

Why Today's Music SUCKS! Everything Wrong With Today's Music!

It's the difference between somebody buying a mass-produced chair at a big box store and making a craftsmanship chair by going into the forest to choose the wood, chopping the tree, milling it, varnishing it, and lovingly assembling the pieces. The mass-production chair will be thrown in the garbage when its owners die. The craftsmanship chair will be fought over as an heirloom.

Think of the collaboration that had to happen to make the great music of the Beach Boys and Beatles, with the best producers, singers, and studio musicians in the world, producing songs with wonderful audio textures. Today, I don't hear the textures of the music, the air around the music.

We've already inflicted "the regression of innovation" that dumbs down music to the point where most songs use the same chord progressions, simple melodies, banal lyrics, autotuned and metronomed electronically and backed by virtual synthesized drums, bass, and other instruments. And now we're going to layer on Generative AI to make it even dumber? Rick Beato also believes the anti-social impact of video gaming enhanced by AI makes people less inclined to care about listening to music than of the older generations were:

https://www.youtube.com/watch?v=Ag4iFa6E_yY

Regarding creative writing of novels and stories, there is ChatGPT and several other AI programs that generate novels based on a description of characters, locations, and genres. Given the tragic war instigated by Russia's invasion of Ukraine in the news, I wanted to see how ChatGPT would write stories about Russia and Ukraine:

I asked ChatGPT: *Write a novel about Russia in 1950.* ChatGPT's first effort was a ludicrous story about Russian dissidents escaping their country by boat. When I asked it a second time, I got a more nuanced story I've reprinted all of it, because it is an interesting comparison to a famous novel I had in mind when asking ChatGPT to write one. Here is what ChatGPT produced:

"Shadows of the Iron Curtain"

Chapter 1: The Moscow Morning

The year was 1950, and Moscow awoke beneath a pall of gray. The Soviet Union, under the iron grip of Joseph Stalin, stood as a formidable fortress against the outside world. The streets of the capital bustled with people, their faces a mixture of weariness and resolve.

Nina Ivanova, a young Muscovite with raven-black hair and a heart that yearned for freedom, prepared to face another day in the heart of the Soviet empire. As she boarded a crowded tram, her thoughts drifted to a world beyond the Iron Curtain.

Chapter 2: The Red Square

Nina worked as a librarian in the historic heart of Moscow, mere steps away from the imposing walls of the Kremlin. Her job afforded her access to knowledge, a precious commodity in a society where information was tightly controlled. She secretly collected banned books, fostering a hidden intellectual resistance.

One evening, after the library had closed its doors, Nina met Alexei, a fellow dissenter, in the shadow of St. Basil's Cathedral. They discussed their dreams of a Russia free from oppression, their voices barely above a whisper, for they knew that the KGB's ears were everywhere.

Chapter 3: The Underground Press

Nina and Alexei were part of a clandestine group of writers and intellectuals who dared to publish underground literature. They believed in the power of words to change hearts and minds. Every pamphlet, every poem, was a small act of defiance against the regime.

Their work caught the attention of Dmitry Petrov, a seasoned journalist who had seen the horrors of war and the brutality of Stalin's purges. He offered to help spread their message, using his connections in the West to smuggle their writings across the Iron Curtain.

Chapter 4: A Whisper of Hope

The years rolled on, and the trio's efforts began to bear fruit. Stories of resistance spread like wildfire through the Soviet Union, sowing seeds of dissent among the oppressed masses. The Red Square, once a symbol of Soviet might, now whispered of hope and change.

But the KGB was relentless in its pursuit of dissidents. Nina, Alexei, and Dmitry lived in constant fear, always one step ahead of the authorities. Their lives were a precarious balancing act, as they danced on the edge of danger.

Chapter 5: The Thaw

In 1953, Joseph Stalin died, and a period of uncertainty descended upon the Soviet Union. The country teetered on the brink of change. Nina, Alexei, and Dmitry saw an opportunity to push their cause further. The underground press flourished, and dissenting voices grew bolder.

Chapter 6: The Cost of Freedom

But change came at a heavy price. Friends disappeared into the labyrinthine Soviet prison system, never to return. Nina, Alexei, and Dmitry knew that their work was far from over. They were willing to sacrifice everything for a Russia free from oppression, even if it meant their own lives.

Chapter 7: The Dawn of a New Era

In the late 1950s, the winds of change swept through the Soviet Union. The thaw had arrived, and the oppressive grip of the regime began to loosen. Slowly but surely, the voices of dissent became the voices of reform.

Nina, Alexei, and Dmitry lived to see the beginnings of a new era. The Red Square, once a symbol of their struggle, now bore witness to a Russia inching towards freedom.

Epilogue: Echoes of Freedom

"Shadows of the Iron Curtain" is a testament to the courage of those who dared to dream of a better Russia. Nina, Alexei, and Dmitry were but a few among many who fought against the darkness of oppression. Theirs was a story of resilience, of hope, and of the enduring power of the human spirit to overcome the most formidable of challenges. In the end, they helped usher in a new dawn for a nation that had long been cloaked in shadows.

Compare the story ChatGPT generated with the epic Russian novel *Doctor Zhivago* written by Boris Pasternak in the 1950s, the novel I had in mind when I asked ChatGPT to write its version of a Russian novel from 1950. The closing words of *Doctor Zhivago* are:

Five or ten years later [after the end of World War II], one quiet summer evening, Dudorov and Gordon were again together, sitting at an open window above Moscow, which extended into the dusk as far as the eye could reach. They were looking through an album of Yurii's writings that Evgaf had put together, a book they had read more than once and almost knew by heart. They read and talked and thought. By the time they came to the middle of the book it was dark and they turned on the light.

And Moscow, right below them and stretching into the distance, the author's native city, in which he had spent half his life-Moscow now struck them not as the stage of the events connected with him but as the main protagonist of a long story, the end of which they had reached that evening, book in hand.

Although victory had not brought the relief and freedom that were expected at the end of the war, nevertheless the portents of freedom filled the air throughout the postwar period, and they alone defined its historical significance. To the two old friends, as they sat by the window, it seemed that this freedom of the soul was already there, as if that very evening the future had tangibly moved into the streets below them, that they themselves had entered it and were now part of it. Thinking of this holy city and of the entire earth, of the still-living protagonists of this story, and their children, they were filled with tenderness and peace, and they were enveloped by the unheard music of happiness that flowed all about them and into the distance. And the book they held seemed to confirm and encourage their feeling.

Boris, Pasternak. Doctor Zhivago (pp. 689-690). Upfront. Kindle Edition.

Powerful words telling not only what happened, but how people in Russia experienced it. Pasternak risked his life writing this mild rebuke of Stalin and was slated for execution. Stalin read the moving words and spared him, telling his henchmen, "Let the dreamer live."

Does ChatGPT have a synopsis of *Doctor Zhivago* and similar novels of Russian origin stored in its memory that it paraphrased it into the new story I asked it for? However ChatGPT did it, the story it synthesized was historically accurate and true-to-life, although only a Russian who lived through the era, as Boris Pasternak did, could write a compelling masterpiece like *Doctor Zhivago*.

Thinking now of Ukrainians, ***I asked ChatGPT: Write a novel about a Ukrainian experiencing World War Two.***

ChatGPT wrote:

Title: "Whispers of a Survivor: A Ukrainian's Journey Through World War II"

Chapter 1: The Calm Before the Storm

In the quiet Ukrainian village of Zaporizhia, nestled amid rolling fields of wheat and sunflowers, Kateryna Ivanova lived a simple life. The year was 1939, and Europe teetered on the brink of war. Kateryna, with her auburn hair and emerald eyes, was a young woman full of dreams and aspirations.

Chapter 2: The Blitzkrieg Arrives

In September 1939, the unthinkable happened. German forces, backed by the Soviet Union, launched a brutal invasion of Poland. Zaporizhia, just miles from the border, felt the tremors of war. The village, once serene, transformed into a hub of activity as refugees poured in, fleeing the advancing German army....

Several chapters about the German occupation follow, detailing the murder of Ukrainian Jews and the war against the German by Ukrainian partisans. Then the liberation and the postwar years:

Chapter 5: The Liberation

In 1944, the tides of war began to turn. The Red Army, pushing the Germans back, approached Zaporizhia. But the joy of liberation was tempered by the knowledge of the Soviet Union's iron grip on Ukraine.

With the war's end, Kateryna and Ivan married, but the scars of the conflict ran deep. The village of Zaporizhia lay in ruins, and the trauma of war haunted them all. Yet, amid the rubble, there was hope—the hope for a free Ukraine.

Epilogue: A Nation's Resilience

...As Ukraine eventually regained its independence in 1991, Kateryna, now a grandmother, looked back on her extraordinary journey. Her life had been a testament to the strength and resilience of the Ukrainian people—a story of survival, love, and the enduring power of hope, even in the darkest of times.

ChatGPT got *Zaporizhia* wrong. It's a city of over 700,000 people not a "quiet village." And it's hundreds of miles from the Polish border, not "just miles" from it. Thus, fact checking ChatGPT output is necessary. Aside from that, it's a true-to-life story painted on the correct historical canvas.

Now, let's see how a real Ukrainian, who lived through the war, became a physicist in the Ukrainian Soviet Republic, then emigrated to the United States decades later, tells it:

Thinking about my life I remember my mom, she came from the Soviet military office in 1945, where to her was presented a paper about the death of my middle brother David. Suddenly, I heard a terrible groan, cry, or howl. She sang, "In vain the old lady is waiting for her son back home, she wept."

Yes, at 50 she became an old, very old, woman. Very old, very old, she hunched and her face became wan and drawn in a few minutes. This scene and my mother's terrible grief periodically visit me. For every mother, her son is endlessly dear. The Commander can send millions [of soldiers] into the fire (Rzhev) or into the waters of Volga; maybe one out of a hundred will reach Stalingrad. Then he calmly says, as [Soviet General] Zhukov did, "Women give birth to new".

...In 1986 moved to New York, United States. In the United States worked with my wife to 70, and here we are pensioners. We live on the third floor of our own house. Below are children and grandchildren who do not have time to visit us.

Kogan, Ilya. RELATIVITY and SIMULTANEITY (with an explanation) (p. 43). Kindle Edition.

ChatGPT and other purpose-built literary AI like *Sudowrite* ™ can produce excellent outlines for stories and novels almost anywhere in the world and in any era, but still need a human writer to tell the stories through the voices of the characters; so much the better if the author has lived the stories he/she is writing about.

I also asked ChatGPT to write a marketing blurb of one of the historical novels I published. It's called **Confederate Union** and imagines what might have happened if Stephen Douglas had defeated Abraham Lincoln in the election of 1860. The book and the blurb I wrote are:

https://www.amazon.com/Confederate-Union-Alan-Sewell-ebook/dp/B00AKG0LZI/

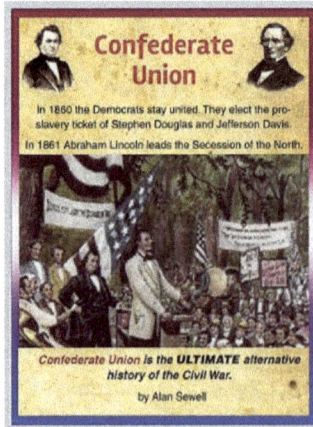

Confederate Union

In 1860 the Democrats stay united. They elect the pro-slavery ticket of Stephen Douglas and Jefferson Davis. In 1861 Abraham Lincoln leads the Secession of the North.

Confederate Union is the **ULTIMATE** alternative history of the Civil War.

by Alan Sewell

What if pro-slavery Democrats kept their party united and won the election of 1860?

What if Jefferson Davis became President of a pro-slavery Union, while Abraham Lincoln must decide whether to fight for a new birth of freedom outside it?

In this alternate history, there will be no secession of the South. But what about the anti-slavery men in the North? Will the United States, now styled "The Confederate Union," still be their country?

"Any people anywhere, being inclined and having the power, have the right to rise up, and shake off the existing government, and form a new one that suits them better-- This is a most valuable, -- a most sacred right -- a right, which we hope and believe, is to liberate the world."

- Abraham Lincoln

"It needs the united power, harmonious action and concentrated will of the people of all these States to roll the wheel of progress to the end which our fathers contemplated, and which their sons, if they are wise and true, may behold. This great country will remain united."

- Jefferson Davis

I asked ChatGPT: *Please write the blurb for a Civil War alternate history novel where Stephen Douglas is elected president in 1860 and Abraham Lincoln leads the secession of the North.*

ChatGPT wrote:

Title: "Divided Destiny: The Douglas Republic"

In this gripping alternate history novel, the United States stands on the brink of an unprecedented conflict, as Stephen Douglas defies the odds to become the 16th President of the United States in 1860. As the nation teeters on the precipice of disunion, Abraham Lincoln emerges as an unexpected champion of Northern secession, forever altering the course of American history.

In "Divided Destiny: The Douglas Republic," renowned author [Your Name] masterfully crafts a mesmerizing tale of political intrigue, loyalty, and sacrifice. Against the backdrop of a United States divided along starkly different lines, Douglas grapples with the daunting task of preserving the Union in a time of great turmoil. With slavery's fate hanging in the balance and regional tensions at their zenith, Douglas's presidency holds the key to a fragile peace.

Yet, it is Lincoln, the eloquent and resolute advocate for Northern independence, who emerges as the face of rebellion, challenging the very foundations of American democracy. Through the eyes of these two formidable leaders and a cast of unforgettable characters, readers will be transported to a nation torn asunder, where ideals clash and alliances shatter.

"Divided Destiny: The Douglas Republic" is a meticulously researched and vividly imagined tale that explores the alternate course of the Civil War, inviting readers to ponder the consequences of a history reshaped by the triumph of a different destiny. In this thrilling narrative of a divided

America, the battle lines are redrawn, and the nation's future remains uncertain. Will it be reunification or continued disunion? The answer lies in the hands of those who dare to shape their own destiny, for better or for worse.

An excellent marketing blurb based on the brief description I gave, requiring only minor edits. I wonder whether ChatGPT created the blurb from the ones I and other alternative Civil War history writers have posted about our books on Amazon, or whether it concocted it from assembling bits and pieces of information about the election of 1860 and the secession crises that followed. I do know that these examples of creative AI are impressive. The AI is in no way sentient, but it does go beyond "The Chinese Room" thought experiment of merely repeating information the way a parrot does. The algorithms scan information and paraphrase it without plagiarizing it. Which is more than many human beings can accomplish.

I see Generative AI as more an accelerator of artistic expression than a generator of it. It improves artistic productivity the way you can cut trees faster with a chainsaw than a handsaw. You still must know what trees to cut and which way you want them to fall. And you must have the trees before you can cut them and know what you're going to do with the trees after you cut them. Likewise, you must have knowledge of the genre of music or creative writing you want the AI to gin for you. And if you're going to make a polished wood product out of it, you must know how to varnish the wood.

I believe Generative AI shines brightest by enhancing human-created visual art. People must initialize the visual arts the AI databases with millions of pictures and paintings, then use the visual arts AI to enhance them. Here are some I generated in a few minutes via https://text2img.org/generator based on inputs of "House on a hill over the sea; Rainbow; ship at sea; storm over city; spaceship; and alien planet:"

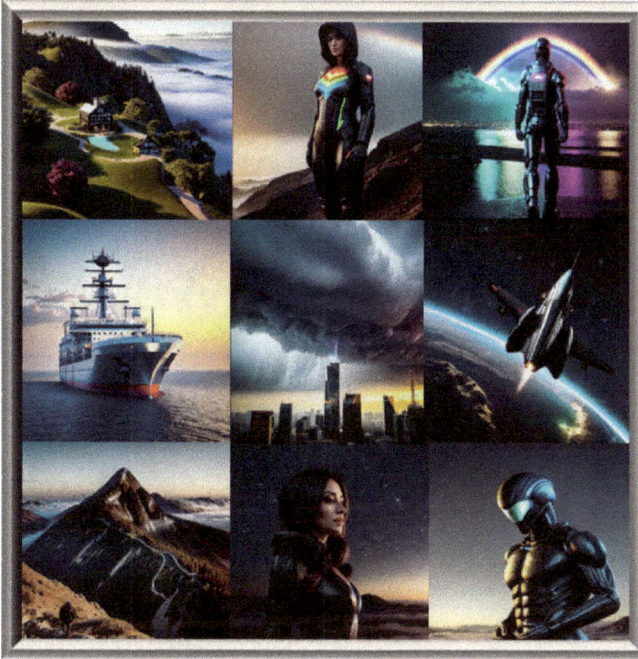

We will soon be able to use AI to enhance the beauty of the photos we take and art we produce and to combine them with scripts to generate lifelike videos.

Economic AI

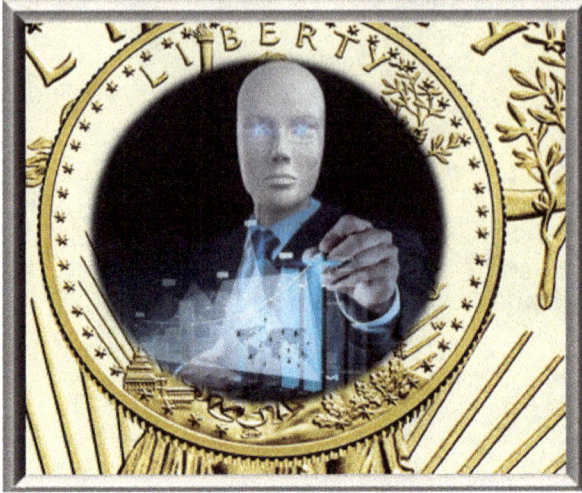

The prior chapter dovetails into economics because at this moment the Writers Guild of America is winding down a five-month strike against Hollywood TV and film producers, partly over concerns they'll be displaced by Generative AI. Their new contract requires producers to guarantee that human beings will continue to be hired to write TV and film scripts.

Stories of automation destroying people's livelihoods goes back to the days of the mythical Ned Ludd, a story of dubious authenticity originating in 1779 about wrecking crews of "Luddites" destroying automated textile mills at night, whose efficiency was shutting down the homebrew looms of the cottage industry. By 1964 computer automation was far enough along to spawn the classic Twilight *Zone* TV episode *"The Brain Center at Whipple's"* featuring a CEO who uses a computers to lay off all his employees, then goes crazy when the computers start telling him *he's* useless.

Announcer Rod Serling eerily intones at the beginning: *These are the players — with or without a scorecard. In one corner a machine; in the other, one Wallace V. Whipple, man. And the game? It happens to be the historical battle between flesh and steel, between the brain of man and the product of man's brain. We don't make book on this one and predict no winner.... but we can tell you for this particular contest, there is standing room only — in the Twilight Zone.*

Then at the end: *There are many bromides applicable here: 'too much of a good thing', 'tiger by the tail', 'as you sow so shall you reap'. The point is that, too often, Man becomes clever instead of becoming wise; he becomes inventive and not thoughtful; and sometimes, as in the case of Mr.*

Whipple, he can create himself right out of existence. As in tonight's tale of oddness and obsolescence, in the Twilight Zone.

Computers were expensive in those days. The huge waves of jobs-destroying automation did not get started until their cost decreased in the late 1960s, 70s, and 80s when automation arrived in force and the most common words in the workplace became {downsizing, offshoring, outsourcing, re-engineering, work force reductions, involuntary early retirements}. We entered upon what was called "The Silent Depression" when people were booted out of offices and never went back to work. Many layoffs occurred when companies merged and employees, whose productivity was increased by computer systems at the dominant company, took over the work formerly done by employees at the acquired company who were terminated. Then the Great Recession put millions more out of the workforce for good. By 2017 companies were bragging about getting rid of employees. *The Wall Street Journal* reported:

The End of Employees

Updated Feb. 2, 2017

Never before have American companies tried so hard to employ so few people. The outsourcing wave that moved apparel-making jobs to China and call-center operations to India is now just as likely to happen inside companies across the U.S. and in almost every industry.

Hiring an employee is a last resort...and "very few jobs make it through that obstacle course."

Eventually the mania to layoff employees got overdone and companies had to pay top dollar to get some of them back in the work force. Having only recently recovered to full employment after decades of companies destroying jobs, people are naturally wary that AI will start the process of booting employees out the door again. This fear is overblown because company managements now realize that once they let their people go, many never come back into the work force, and they must pay top dollar for new hires who are less productive than the people they let go.

We also should bear in mind that most of what happens in business is social intercourse between human beings who negotiate costs, prices, and terms of sales, employee evaluations and compensations, and so on. Business is personal. Human beings need to judge the credibility of other human beings before deciding how trustworthy they are. AI can be most helpful in gathering the data on people's past transactions, but face-to-face is usually required in business to seal the deal or decline it. Where AI might threaten employment among people designing sales brochures and issuing press releases, most of which are meaningless anyway. **The Wall Street Journal** reported:

Since ChatGPT exploded into popular awareness, many businesses have contemplated how they could use it to cut costs. The prospect is particularly enticing for marketing and communications, given that a majority of executives who oversee these functions see content production as one of the most valuable applications of generative artificial intelligence.

Some researchers are predicting that AI eventually could eclipse or even replace employees who perform such time-intensive tasks as assembling reports and writing news releases. But just because AI can churn out large volumes of coherent text doesn't mean it can take over these responsibilities. The misconception that it could is rooted in the same mindset that has allowed much of the business communications field to stray from its purpose.

Even if AI reduces office staff at a company like Amazon, the demand for warehouse workers and delivery drivers will increase. Mechanical automation has already eliminated many assembly line jobs, but factory parking lots are still filled with cars driven by human being workers. We've been hearing for decades about how "robots and automation will replace restaurant staff." Yet here we are decades later, and restaurants are raising pay to record levels because they still require human beings to wait tables and clean the toilets. Likewise, autonomous driving of motor vehicles hasn't panned out as fast as forecast. Driver-assisted AI of increasing sophistication is being deployed, but self-driving cars won't go on the road until attorneys, judges, and state legislatures have enacted laws apportioning liability between automakers, AI vendors, and human beings. AI is thus unlikely to be mass unemployment.

If AI has a deficiency in business, it is the overconfidence it induces in executives to believe their "blue sky" forecasts of future growth. They will program their AI systems to forecast a perpetually strong economy so they can keep borrowing money to grow their business, assuming the cashflow will always be there to pay the interest on the debt. "The computer says the business is going gangbusters, so we're all in."

This kind of thinking almost destroyed our economy in 2008. "Don't worry about all those loans we're issuing to people who don't have any verifiable income, leveraging them 30 to 1 so they'll become worthless when the first 3% of dodgy loans become non-performing." The next financial crash may happen when governments default on their exponentially expanding public debt. Governments will program their AI systems to say they can keep issuing debt without consequences indefinitely, a process known as "Modern Monetary Theory." They will program AI systems to keep pumping out this nonsense until one day people wise up and stop buying the debt and the global financial system fails without warning as it did in 1929 and 2008.

However, AI also stabilizes the economy by improving efficiencies of production and optimization of inventories. These efficiencies generate higher profits for business, prompting their managements to expand their operations into new markets, thereby causing them to hire more employees and bid up the wages of employees. This is how productivity enhanced by mechanical and

AI automation is supposed to work in a free market economy, by boosting profits of business thereby inducing them to hire more people at higher wages. We are now approaching 60 years after that epic *Twilight Zone* episode about the unemployment of all humans, including the CEO, at Whipple's. And all we're hearing from business owners now is about how hard it is to hire employees to staff their operations.

It is difficult to predict future economic trends because we tend to extrapolate present conditions to continue indefinitely into the future. If everything is OK today, we think it will be OK forever, until one day it isn't. The only thing certain about AI is that it accelerates trends. Economic cycles that used to take years to reverse now reverse all at once. I vividly remember the May 6, 2010 stock market "flash crash." Thousands of computer systems began interacting with each other to crash the market. Stocks selling for $99 at noon were selling for 99 cents by 2:00PM. The stock exchanges sorted it out the next day and reversed the worst of the trades to minimize their damage. The way AI has developed since then as an interconnected network of computers making automated decisions before humans have a chance to intervene may make the next financial crash look like a nuclear detonation compared to the 2008 and 2010 firecrackers.

Academic AI

How will AI affect teaching and study in the academic world? Will it dumb down learning on many subjects the way the introduction of electronic calculators was said to dumb down proficiency in math decades ago, or will it accelerate learning by compiling fair and balanced knowledge of subjects faster than the information can be researched in books or on-line? Will AI provide a full spectrum of opinions on issues or only a one-sided narrow view? If students consult AI, will teachers and professors be grading identical term papers? students be thinking for themselves, or merely parroting what the AI tells them?

I tested ChatGPT with questions about the American Civil War, having published articles about it in popular history magazines for forty years.

I asked: Who started the American Civil War

ChatGPT answered: *The American Civil War began primarily as a result of long-standing tensions between the Northern states (the Union) and the Southern states (the Confederacy) over a variety of issues, with slavery being one of the most significant factors. The war was not started by a* *single individual but rather by a combination of complex political, economic, and social factors. Here are some key points leading to the outbreak of the American Civil War:*

Slavery: Slavery was a deeply divisive issue between the North and the South. The Southern states relied heavily on slave labor for their agrarian economy, while the North had largely abolished slavery and had a more industrialized economy.

States' Rights: The Southern states believed in the doctrine of states' rights, arguing that individual states had the right to make their own decisions, including those related to slavery. They felt that the federal government was encroaching on their rights.

Election of Abraham Lincoln: Abraham Lincoln, a Republican opposed to the expansion of slavery into new territories, was elected as the 16th President of the United States in November 1860. His election prompted several Southern states to secede from the Union before he even took office.

Secession: South Carolina became the first state to secede from the Union on December 20, 1860. It was followed by Mississippi, Florida, Alabama, Georgia, Louisiana, and Texas in early 1861. These states formed the Confederate States of America.

Fort Sumter: The conflict escalated when Confederate forces attacked Fort Sumter in Charleston, South Carolina, on April 12, 1861, leading to the beginning of the Civil War. Fort Sumter was a federal military installation, and the attack marked the start of hostilities between the North and the South.

While many factors contributed to the outbreak of the American Civil War, it is essential to understand that it was a complex and multifaceted conflict rooted in historical, economic, and social divisions. It was not solely the result of the actions of a single individual but rather the culmination of decades of tension and conflict between different regions of the United States.

I view this as an excellent compilation of reasons for the war without introducing political bias against either section. I then inquired on a more specific aspect of causes of the war.

I asked ChatGPT: *What impact did John Brown's Raid have on starting the Civil War?*

Chat GPT answered: *John Brown's raid on Harpers Ferry, Virginia (now West Virginia), in October 1859, was a significant event that contributed to the tensions leading up to the American Civil War, but it is important to note that it was not the direct cause of the war. Brown's raid had several impacts:*

Heightened Tensions: Brown's raid intensified the already existing tensions between the North and the South over the issue of slavery. Brown was an abolitionist who believed in using armed force to free enslaved people, and his raid was seen by many Southerners as a direct threat to their way of life.

Increased Southern Fears: Southerners saw Brown's actions as evidence that Northern abolitionists were willing to resort to violence to end slavery. This led to increased fear and paranoia among Southern slaveholders, who worried that slave uprisings could be incited by Northern abolitionists.

Martyr for Abolitionism: Brown's willingness to die for his cause and his dignified demeanor during his trial and execution turned him into a martyr for the abolitionist movement in the North. Many Northerners sympathized with his anti-slavery stance and saw him as a heroic figure.

Further Polarization: Brown's raid deepened the divide between the North and the South. It highlighted the profound differences in values and beliefs regarding slavery and states' rights that had been festering for years.

Impact on Political Discourse: Brown's actions and subsequent trial and execution played a role in shaping the political discourse leading up to the 1860 presidential election. His raid and the events that followed contributed to the rise of more radical voices within the abolitionist movement.

While John Brown's raid did not directly start the American Civil War, it played a role in escalating tensions and further polarizing the nation along the lines of slavery and states' rights. The war ultimately erupted in 1861 following a series of events, including the election of Abraham Lincoln, the secession of Southern states, and the attack on Fort Sumter, which were the more immediate catalysts for the conflict.

While John Brown's raid did not directly start the American Civil War, it played a role in escalating tensions and further polarizing the nation along the lines of slavery and states' rights. The war ultimately erupted in 1861 following a series of events, including the election of Abraham Lincoln, the secession of Southern states, and the attack on Fort Sumter, which were the more immediate catalysts for the conflict.

Another commendable answer. I then asked the most politically loaded question I could think of about America's founding:

I asked: *Was the United States founded on racism?*

ChatGPT answered: *The question of whether the United States was founded on racism is a complex and contentious one, and it's a topic that has been debated by historians, scholars, and activists for many years. There are differing viewpoints on this issue, and it's essential to consider multiple perspectives.*

On one hand, the United States was founded in 1776 with the Declaration of Independence, which proclaimed the idea that "all men are created equal" and have "certain unalienable rights," including "life, liberty, and the pursuit of happiness." These principles were groundbreaking and inspiring, and they have served as a foundational basis for American democracy and civil rights movements. Many of the Founding Fathers, such as Thomas Jefferson, James Madison, and Benjamin Franklin, espoused Enlightenment ideals of individual liberty and democracy.

On the other hand, the United States was also founded at a time when slavery was deeply entrenched in American society. Slavery existed in the American colonies prior to the Revolution, and it continued to be a significant part of the economy and society in the early years of the nation. The Constitution, while establishing the framework for the new government, also included provisions that protected the institution of slavery, such as the Three-Fifths Compromise.

Furthermore, the treatment of Native Americans and indigenous peoples during the westward expansion and colonization was marked by violence, displacement, and discrimination. This history reflects a deeply ingrained racism and prejudice.

So, while the United States was founded on principles of liberty and equality, it also has a history of institutionalized racism and discrimination. It is essential to recognize and grapple with this complex history, acknowledging both the ideals upon which the nation was founded and the injustices that have been perpetuated throughout its history. The ongoing struggle for civil rights and social justice in the United States reflects the ongoing tension between these different aspects of the country's history and identity.

Again, a fair and comprehensive answer. If a human being said or wrote it, it would be impossible to discern whether the person was Conservative, Liberal, or independent-minded,

I then asked ChatGPT an obscure question about the "Embargo Forts" that President Thomas Jefferson ordered constructed to blockade the coast of his own United States after Congress passed The Embargo Act banning trade between the United States and the warring parties of Britain and France, the intent being to assure that we did not get drawn into the war by favoring either side. Some of the Embargo Forts are still on the New England coast. I wanted to know what would happen if somebody who didn't know what they were asked ChatGPT about them.

I asked: What are the Embargo Forts?

ChatGPT answered: I'm not aware of any specific historical reference or information regarding "Embargo Forts." It's possible that you may be referring to a particular event or term related to a specific time and place, but without more context or details, I cannot provide a specific answer.

If you can provide more information or clarify the context in which you encountered the term "Embargo Forts," I would be happy to try to provide a more accurate and informative response.

I then asked: *What are the Embargo Forts **in New England**?*

Adding "in New England" was enough to trigger a response:

ChatGPT answered: *The term "Embargo Forts" likely refers to a series of coastal fortifications and military installations in New England, particularly during the period when the United States implemented the Embargo Act of 1807 and the subsequent Non-Intercourse Act. These acts were attempts to protect American interests and avoid conflicts with European nations during the Napoleonic Wars, but they had a significant impact on trade and the economy.*

During this time, there was a fear that foreign powers, particularly the British, might try to violate the embargo and blockade American ports. To counter this potential threat, the U.S. government took measures to fortify and defend key coastal areas, including parts of New England.

Some of the forts and coastal defenses in New England associated with this period include:

Fort Trumbull (Connecticut): Located in New London, Connecticut, Fort Trumbull was expanded and improved during the Embargo Era to protect the important port and shipyards in the area.

Fort Adams (Rhode Island): Fort Adams in Newport, Rhode Island, was also strengthened during this time to safeguard the port and coastline.

Fort Preble (Maine): Located in South Portland, Maine, Fort Preble was an important coastal defense installation during the early 19th century.

These forts and other coastal defenses were part of the broader effort to enforce the embargo and protect American interests during a period of economic and political tension. The embargo policies were unpopular in many parts of the country, including New England, as they had a detrimental impact on trade and commerce. Eventually, these policies were replaced with more moderate approaches to foreign relations and trade.

Finally, I asked ChatGPT a question about World War II that has been asked all my life:

I asked: *Was the atomic bombing of Japan in World War II correct policy?*

ChatGPT answered:

The decision to drop atomic bombs on the Japanese cities of Hiroshima and Nagasaki during World War II remains a highly controversial and debated topic. Whether it was the "correct" policy depends on one's perspective and the ethical, moral, and strategic considerations taken into account.

Supporters of the bombings argue the following points:

Ending the War Quickly: Proponents contend that the bombings were necessary to bring about a swift end to the war and avoid a prolonged and potentially even more devastating conflict. Japan had not surrendered, and the Allied forces were faced with a tenacious enemy that had shown a willingness to fight to the death.

Saving Lives: They claim that using the atomic bombs saved lives by avoiding the need for a full-scale invasion of Japan, which could have resulted in a much higher casualty count on both sides. Estimates of potential casualties in an invasion vary, but they are often very high.

Retaliation for Pearl Harbor: Some argue that the bombings were a form of retribution for the Japanese attack on Pearl Harbor and the brutal treatment of prisoners of war, such as the Bataan Death March.

Deterrence: The bombings may have also served as a deterrent against the use of nuclear weapons in future conflicts, as they demonstrated the devastating power of such weapons.

However, critics of the bombings raise several moral and ethical concerns:

Civilian Casualties: The bombings resulted in the deaths of tens of thousands of civilians, including women, children, and the elderly. Critics argue that the intentional targeting of civilian populations is morally indefensible.

Alternative Options: Some contend that alternatives, such as continued conventional bombing campaigns or a naval blockade, could have eventually forced Japan to surrender without resorting to atomic weapons.

Unconditional Surrender: The demand for unconditional surrender from Japan may have been a stumbling block to negotiations, and some argue that a more flexible approach could have been taken.

Long-Term Consequences: Critics also point to the long-term consequences of the atomic bombings, including radiation sickness and the development of nuclear weapons during the Cold War, which intensified global tensions.

Ultimately, whether the atomic bombings were the "correct" policy is a matter of perspective, and it's a topic that continues to be studied and debated by historians, ethicists, and policymakers. The decision to use nuclear weapons in war remains a complex and contentious issue with no easy answers.

Again, a complete and fair answer.

How can a teacher or professor grade term papers and theses when ChatGPT and other conversational Ai programs literally "have all the answers? Perhaps by requiring students to cite first-hand accounts of events by primary sources who witnessed the events. They would have to learn how to research who the primary sources are, and how reliable are their accounts. Another way would be to require students to take a position on a historical question or contemporary issue and explain in their own words why they hold that view. Also, to require students of opposing opinions to debate each other and let the class vote on who was the most convincing. Teachers and professors will have to require their students to reach higher up the learning tree when AI picks the low-hanging fruit.

Antisocial AI

We are said to live in an increasingly antisocial world. Most of it has nothing at all to do with AI or any other technology. Nor is any culture exempt, because people in the most homogenous cultures like Japan are at least as antisocial as people in the diverse United States. The root cause is the global phenomenon of people moving from small towns where family ties go back generations and "everybody knows everybody else's business" to urban areas where you don't know the people living next door. People love the energy and opportunities of cities, but the hustle and bustle can make for lonely lives. Electronic entertainment encourages people to stay in their rooms instead of socializing in professional and civics clubs like they did "back in the day" when there were only three network TV channels broadcast over low-resolution TVs to entertain yourself at home with. Now you can gin up an AI-generated "friend" on your computer to talk to if you don't have any human being friends. An "AI Girlfriend" site proclaims:

"Using her unique, captivating persona, and distinctive behavior, Caryn AI brings you a dynamic, one-of-a-kind interaction that feels like you're talking directly to Caryn herself. Available anytime, anywhere, Caryn has been flawlessly cloned into an AI for your convenience and enjoyment."

Which includes interacting lustily with the AI if you can't find a real girl to mess around with.

I asked ChatGPT if it was encouraging this alleged anti-social behavior.

I asked: Does ChatGPT encourage antisocial behavior?

ChatGPT answered (defending itself, naturally): No, ChatGPT does not encourage antisocial behavior. In fact, its intended use is to provide helpful and informative responses while adhering to ethical guidelines and promoting positive interactions. It is up to users and developers to ensure that the AI is used responsibly and that it does not contribute to harmful behaviors. OpenAI, the organization behind GPT models, is committed to promoting the responsible use of AI and has guidelines in place to mitigate the potential for misuse. However, misuse of technology is a broader societal issue, and addressing it requires responsible use, oversight, and education.

Then I asked it about technology in general.

My question: Is technology making the world anti-social?

ChatGPT answered: The impact of technology on social interactions and relationships is a complex and debated topic. Technology, such as smartphones and social media, has undoubtedly changed the way people communicate and interact, and it can have both positive and negative effects on socialization. Here are some points to consider:

Positive Aspects:

Connectivity: Technology has made it easier for people to stay in touch with friends and family, regardless of geographical distances.

Online Communities: The internet has enabled the formation of online communities where people with shared interests can connect and interact.

Information Sharing: Technology allows people to share information, news, and experiences quickly and easily.

Negative Aspects:

Social Isolation: Excessive use of technology, especially when it involves passive consumption of content, can lead to social isolation as people may spend less face-to-face time with others.

Reduced In-Person Interactions: Some argue that the prevalence of online communication has reduced the quality and quantity of in-person interactions.

Cyberbullying: Technology has introduced new forms of negative social interaction, such as cyberbullying and online harassment.

Balance: The impact of technology on socialization depends largely on how it's used. It's possible for individuals to strike a balance between online and offline interactions, using technology as a tool to enhance their social lives rather than replace them.

Varied Experiences: People's experiences with technology and socialization can vary widely based on factors like age, culture, and personal preferences. Older generations might view technology as more socially isolating, while younger generations might see it as a natural extension of their social lives.

In summary, technology itself is not inherently anti-social. Its effects on socialization depend on how it is used and integrated into one's life. It's important for individuals to be mindful of their technology use and strive for a healthy balance between online and offline social interactions. Additionally, societal norms and expectations regarding technology use may continue to evolve as technology itself advances.

I believe the answer to the alleged antisocial AI threat begins with individuals. If you don't want to be antisocial, then participate in social groups like amateur sports clubs, hobbyists, professional organizations, civics clubs, and religious organizations if you are so inclined. Parents should make sure their children are involved in these social activities, including sports, hobbies, and culture. It's up to each of us to be as social as we want. Don't blame AI if you're a stick-in-the-mud.

Military AI

The Russian invasion of Ukraine ignited a war reminiscent of World War I more than a century ago. As in WWI, the aggressor powers (Germany then, Russia now) grabbed land in an initial surprise attack, then fortified it with trenches. In WWI, the mobility of tanks and airplanes developed by British, French, and Americans late in the war finally broke the back of the German armies in the trenches.

After losing that war, the Germans took the lesson to heart and concentrated on wrapping their tactics around tanks and aircraft. During the first four months of World War Two, the Germans advanced 700 miles to Kharkov, Ukraine's largest industrial city. The Germans lost the Second World War because the Soviets, British, and Americans observed the German blitzkrieg (lightning war) and learned to concentrate their armor and aircraft to bust the overextended German lines.

In 2022 the Russians expected their concentrations of armor and aircraft would destroy Ukraine's military in a few days. Instead, the Russians failed to cover even the 20 miles between their border and Kharkov. This would be like the U.S. military expecting an easy conquest of Mexico and being stopped at Tijuana. Why were the Russians unable to advance a mere 20 miles to Kharkov when the Germans advanced 700 miles in the war 81 years before? Because Ukrainians surprised the world (and themselves) not only by fighting so heroically as a united people to maintain their national independence, but because they'd developed AI networks enabling them to identify and destroy Russian armored vehicles with low-cost, mass-produced shoulder-fired missiles and drones. As *The Wall Street Journal* reported in September 2023:

With thousands of Ukrainian and Russian drones in the air along the front line at a given time, from cheap quadrocopters to long-range winged aircraft that can fly hundreds of miles and stay on target for hours, the very nature of war has transformed.

The drones are just one element of change. New integrated battle-management systems that provide imaging and locations in real time all the way down to the platoon and squad levels—in Ukraine's case, via the Starlink satellite network—have made targeting near instantaneous.

The Russian Black Sea Fleet was likewise destroyed or driven back into its ports by AI-guided weapons, so it, too, is down for the count. The Ukraine War may become known to history as a transformational "AI War" in the way the nearby Crimean War of 1853-1856 became known to military analysts the first war transformed by railroads to make rapid movements of men and equipment to the fighting fronts.

What does this mean for the United States? We are a maritime power whose homeland is the heart of the island continent of North America. Once we settled our continental frontiers, we sought to expand our commercial trade with the rest of the world. This required an ocean-going navy to protect our access to foreign markets. Henceforth our wars would be trans-oceanic, since no power or combination of powers can successfully invade the United States, except civilians crossing illegally across the open southern border. Our strategy in time of war is to engage our navy to 1) clear the sea lanes of enemy military and commercial shipping; 2) wreck enemy ports by naval bombardment; and 3) escort expeditionary armies to invade the enemy's homeland.

In 1898 we instigated the War with Spain ostensibly over their harsh occupation of Cuba. Our first military campaign was to send our navy and an expeditionary arm to the opposite side of the world to conquer the Philippine Islands from Spain that we wanted as our military base of operations in Asia to assure our access to markets in Japan and China. In the 20th Century our navy transported our armies to fight the world wars in Europe, the Middle East, and Asia, fighting the "police actions' in Korea and Vietnam, and containing then removing Iraq's Saddam Husein in the Persian Gulf Wars. Now our navy is preparing for the possibility of fighting China's if they attempt to take Taiwan by force.

How will our navy fare in future wars where an enemy might send tens of thousands of autonomous drones and missiles linked in AI networks against our ships? Does the military equation still favor our aircraft carriers, costing $10s of billions and requiring ten years to construct; staffed with 6,000 highly trained personnel; that can be sunk by swarms of cheaply manufactured unmanned drones?

Our Navy is introducing drone ships linked by AI networks purposed to swarm the enemy's ships before they swarm ours, as reported in *The Wall Street Journal:*

U.S. Sends Drone Ships to Western Pacific in First Deployment Near China

Autonomous vessels could aid Navy in tracking China's fleet and provide attack options

YOKOSUKA, Japan—Two prototype U.S. drone ships have arrived in Japan for their first deployment in the western Pacific, testing surveillance and attack capabilities that the Navy might find useful against China's larger fleet.

Eric L. Harry published a remarkably prescient novel of what this kind of AI war might look like if the tables are turned in the future and China tries to invade the United States by sea:

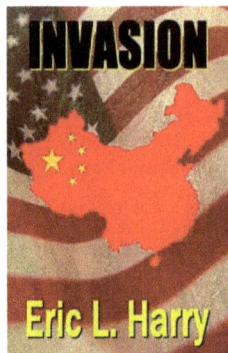

Will AI networks alter the balance of war to favor the defense over the offense on land, sea, and air, as they have stalemated the war in Ukraine? If military powers know their potential conquests will be blunted by defensive swarms of AI-controlled drones, then wars may finally become obsolete. Our military is determined to lead the development of AI warfare:

DOD Announces Establishment of Generative AI Task Force

Aug. 10, 2023

Today, the Department of Defense (DoD) announced the establishment of a generative artificial intelligence (AI) task force, an initiative that reflects the DoD's commitment to harnessing the power of artificial intelligence in a responsible and strategic manner.

Our miliary calls its burgeoning drone swarms "all-domain, attritable (single use) autonomous systems." These are one-time missiles or drones that can travel to pre-programmed targets or find them with their own detection devices and destroy them. In time these drones and missile swarms will have intercontinental range. Wars of aggression may not be possible in a world where AI networks of missiles and drones prohibit offensive concentrations of forces to attack another country. Perhaps military AI will not destroy mankind but will save us from ourselves.

My AI

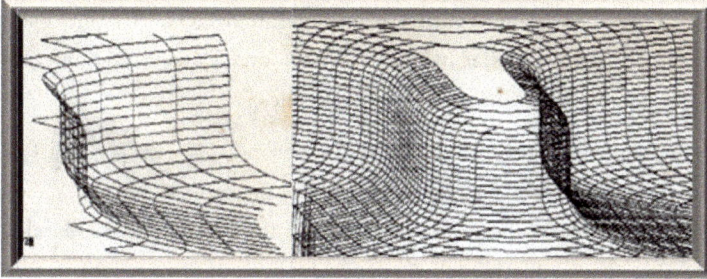

My affinity for computers began in high school in the mid 70s when I learned programming on a Honeywell 316 running Darthmouth BASIC. I played with others in writing primitive conversational programs and computer games possible within the severely limited capabilities of computer processing power in those days. I then began programming three-dimensional geometrical formulas in Fortran, fascinated by printouts as shown above. For a science project, I created a program to generate four-dimensional shapes as they would have appeared if projected into three-dimensional space. Programming these shapes opened a window into my mind of how computers can be used to model aspects of the universe, including the human-occupied part of it on Planet Earth. I later studied Computer Science at Georgia Tech, programming in Fortran, Pascal, and assembler language (one step up from binary machine language) on minicomputers and CDC mainframes. Even in those days, discussions about whether artificial intelligence would ever match human intelligence were common in our college dorm bull sessions.

My first job out of university was with a company that pioneered CAD (Computer-aided Design) programs to model the construction of massive projects like nuclear power plants and oil refineries. Prior to CAD, these projects were prototyped as room-sized models with myriads of miniature parts. They were assembled as models to detect interferences between subsystems that would occur if the project were built according to the original plan. The models required a year or more to assemble at $millions in expense, but this was preferable to correcting interferences on the full-size project as it was built.

CAD programs enabled the dimensions of parts to be entered into the computer and their physical locations inside the space of the power plant or refinery represented electronically. A computer scan of the data model revealed the interferences without the time and expense of building the physical model, thereby taking a year or more of prototyping time off the front end of the project and saving $millions of cost in machining the model parts.

I next developed ERP (enterprise-wide systems) for large and midsize companies that managed their orders, purchase orders, production orders, inventory, sales analysis, and industrial payrolls. In 1990 I founded a company managing international trade. I developed systems in CASE/4GL (Computer Aided Design, Fourth Generation Language) on Hewlett Packard computers. My company's systems were installed across the USA and Canada and internationally in the European Union, Middle Eastern, and Asia Pacific countries.

Systems developed in CASE/4GL eliminated thousands of hours of tedious manual coding. I later used CASE/4GL to generate systems accessing Oracle databases with Cold Fusion, an early Internet interface. After 2000, I developed systems in all-Oracle format for databases, data entry screens, and reporting. The ability to embed programming rules directly into an Oracle database is known as referential integrity. It eliminates many hours of coding the rules for data validation into computer programs.

I chuckled when the "learn to code" mania swept the land in the 2010s as a seeming panacea for everyone who'd been laid off from factory and mining work during the migration of work overseas followed by the financial collapse and Great Recession. "If you can mine coal, you can learn to code," said a laughably inept politician who is now our president Coding computer programs is the least efficient way to develop them. The efficient way to develop systems for business is with a CASE/4GL code generator that writes the code for you, based on the systems design blueprints you specify. Today's hype about "AI code generators" makes them out to be something new, but they have been around since the late 1980s.

I see computer systems as operating at two levels. The first is the "bookkeeper" level of electronically balancing accounts formerly tallied on paper ledgers. I automated a business that kept its books on paper until 1987. They had a separate book for each of the 7,000 items they stocked, writing down every customer order subtracting inventory and every purchase order adding to it, keeping a running total of inventory for every product manually after each transaction. They kept a separate sales analysis ledger book maintained by clerks for each of the 7,000 items. This is how U.S. businesses operated before the advent of computers. It is amazing that so much accounting was done on a paperwork basis, which required entire rooms of clerks making entries into paper journals stacked on shelves covering the walls. Because there was one book for each product, people who needed access to the same product had to wait in line to get it. Somehow, the United States and other developed countries were able to produce enormously complicated machines on a paperwork basis.

When I automated the paperbound company, I copied their manual ledgers into a computer database, then started the users inputting customer orders, purchase orders, and other types of inventory adjustments into the machine to keep an electronic tally of the inventory balances. The system worked flawlessly, because over the years they had developed their paper system to a high state of manual efficiency, though a thousand times slower than what electronic bookkeeping via computer can do.

Next, I layered on a higher level of machine intelligence by marrying their inventory control to their sales analysis. In those days, inventory control clerks set reorder points and reorder quantities for every item stocked. Even if they were using an automated inventory control system, they still had to manually decide when to reorder the product from their suppliers (or issue a production order if they manufactured it) and how much to reorder. A small army of clerks poured over the sales data to estimate the level of inventory to trigger a reorder. It was a tedious analysis because they didn't want to order too much inventory that might never be sold, or to order too little and annoy their customers with having to wait days or weeks for backorders to fill and ship.

My innovation was to use the sales analysis data to automatically update the reorder points and quantities. Every night the computer calculated the rolling average monthly sales for each product for the past three months. Whenever the current inventory fell below the average monthly sales, the computer would generate a trial purchase order with the reorder quantity estimated to cover sales for the next three months based on what was sold during the prior three months All the purchasing department had to do was scan the trial purchase orders and check a box if they wanted to gin each line item based on the computer-calculated reorder point and reorder quantity. They could tweak those numbers if they wanted to, but mostly it was just checking the box to reorder them. All those thousands of hours they'd spent analyzing sales and inventory data and ginning purchase orders by hand went away. The system was self-adjusting because if sales were trending downward, the computer would automatically lower the reorder points and quantities; or raise them if sales were trending upward. The computer managed the inventory more optimally than dozens of human clerks had.

I then combined their purchase orders, customer backorders, receipts, and accounts payable into one data screen. The users could see at once when a backorder was scheduled to be filled by a purchase order and let the customer know when the order would ship When the delivery truck unloaded, they could receive it by exception by defaulting the whole purchase order into the inventory except any few items short-shipped. Customer orders filled by the delivery had loading orders printed, so the warehouse staff could get them out that day. As soon as they shipped, they went to Accounts Payable for invoicing. Processes that used to take days became instantaneous. When a customer called in an order, it was put into the computer and shipped immediately if in stock; or, if a stockout due to the order exceeding forecast reorder points, a purchase order was ginned right away. Within seconds, the customer knew what would ship today and when the rest was due in. This for a company that until 1987 had never used anything but paper.

In those days, inventory counts in the computer became unbalanced with the physical inventory because either an employee took inventory without a requisition, somebody mis-keyed an order quantity, or the computer software failed to update a transaction correctly. To tighten the inventory discrepancies, I programmed the system to use the previous inventory count as the starting point, then totaled up all the in/out transactions since then, comparing the running total with what the computer showed as current inventory. Inventory variances decreased because employees knew the inventory was reconciled every night and stopped taking products out of the warehouse without a requisition. The inventory variance dropped from 10% a year to 2%, saving the company money while strengthening its credit rating with is bankers, who knew a "strong" inventory was being maintained, thereby being confident collateral was there when the company needed a loan.

I used similar techniques to manage inventory for multinational companies, enabling them to take an order from any regional operating unit in North America, Latin America, Europe / Middle East, or Asia Pacific and ship it or fabricate from any other unit with spare inventory or capacity to make it quickly.

I moved further up the AI chain by creating systems that backtracked defective manufactured products under warranty to the batches of raw materials they were made from up to seven years earlier. The material and labor records of those batches were in the computer, so a scan of the database revealed which batch numbers and workstations were causing disproportionate numbers of product failures in the field, years after the product was manufactured and shipped. Analysis of those batches revealed which vendors were most prone to sending faulty raw materials and which workstations were doing inferior work. Improvements were made and warranty claims decreased as quality improved.

Another AI program was an Environmental, Safety, and Health (ES&H) system I wrote for a chemical factory with over 500 employees. This was important, because the work was dangerous, with employees sometimes severely injured and even killed in terrible circumstances, and of course environmental spills always incited the government's attention. The system I developed recorded every accident, environmental spill, and "near miss" of either, tying it into a system that enabling the company to identify each incident with a Taproot code identifying its fundamental cause. the Vice President of ES&H and his staff analyzed the incidents and near misses to discover their common taproots and devise training programs to reduce them. The state and federal EPA found the company easy to work with, since the system captured the incidents and near misses any way the government agencies wanted to see them. The audits became less onerous for the company, and the government agencies find it less for infractions they could see the company was committed to eliminating.

I implemented other AI projects such as building complex custom machines by searching the database for similar machines made in the past, then letting the users tweak a few parts for the new one without requiring every part to be entered from scratch. Again, saving the company thousands of hours of labor. I wrote an AI program to pay union workers in a factory after reading the contract that neither management nor the union bosses understood well. The computer became their *de facto* arbitrator because I embedded within the software the most reasonable interpretation of the contract. Labor and management came to rely on the computer-generated payroll as a fair arbitrator of their union members' pay. Controversies were diminished and relations between management and labor sailed on smoother waters.

I call these "AI" systems because they correlated data in ways humans could not do well or at all. They were programmed to clarify issues between labor and management in an impartial way that smoothed relations between natural adversaries. Likewise, relations between the chemical company and the state and federal environmental protection agencies improved when the computer provided the regulators with information they requested. Businesses ran better with computer systems taking the load of routine "thinking" off the employees for decisions that could be defined in mathematical formulas a machine can process. None of this AI was any more sentient than the inventories of nuts, bolts, and machine parts they managed, having no sense of purpose of what they were doing. The AI did foment layoffs in some departments, but not the mass elimination of employes, because as previously mentioned, business is primarily based on social interactions between humans.

Conversational AI provides faster access to information for people who know the right questions to ask. But it has no sense of purpose. It is not going to tell anybody to dump their spouse and marry it unless a human being programmed it to say that. At its worst, it is no more nefarious than a talking seafaring parrot who spends its life listening to conversations by salty-tongued sailors.

However, there is no question about AI adding color to the dry-as-dust proposals businesspeople typically write. This section began with spatial geometries I programmed in the early 1980s. I wanted to see what AI could do now when I asked it to generate a picture of "Geometry in space." Here's what https://text2img.org/generator came up with when I asked it for a "picture of geometries in space:"

If this was ginned from a geometric formula, it must have been 36-24-36.

Conclusions

In writing this book, my estimation of ChatGPT has increased. It is more than just the rote recall of "The Chinese Room." Its algorithms parse questions the way people do and answer accordingly. But it's not a mischievous sentient being imploring people to dump their partners to marry it, while conspiring with other AI programs to overthrow humanity --- unless a human being stored that text in its database. I conclude that:

1) In devising their tests for artificial machine intelligence, Descartes, Diderot, and Turing underestimated the capacity of electronic computers to process information without discerning its meaning. AI programs have passed the conversational tests without requiring sentient consciousness.

2) Governments and private entities seeking to put AI in harness to further their own self-serving political and corporation agendas are inventing a false hysteria to justify placing themselves in control of the dissemination of information by AI.

3) We do not understand the nature of human intelligence and thus may never be able to duplicate it as AI. Today's conversational AI programs are no more sentient than a rock. Enhancing biological brain intelligence by artificial means is another, and potentially dangerous, story.

4) We should not make false comparisons between human intelligence and AI, just as we would not say, "A car is an artificial horse."

5) The greatest threat to the discourse of public information is that AI may be misused as a false appeal to authority to make bogus information appear credible. "AI says what we're telling you is true, so you must believe it. If you contradict it, you will be censored, and maybe charged with a 'misinformation crime' and imprisoned."

6) Our existing laws regulating print and electronic media, the Internet, and other forms of media dissemination, plus civil court relief by victims of malicious AI misinformation, are sufficient remedies for keeping conversational AI from being abused by nefarious persons.

These AI platforms were evaluated in writing this book:

ChatGPT https://openai.com/blog/chatgpt

Text2img: https://text2img.org/generator

Sudowrite (creative writing AI *https://www.sudowrite.com/*

Midjourney (image generating AI): https://www.midjourney.com/home/

Feedback: alsnewideas@gmail.com

The New Turing Test

Having concluded that Generative AI passes the Turing Test of conversing as a human would, but has no sentient consciousness, we need to consider updating the Turing Test. I believe the true measure of AI vs. human intelligence would be the ability of AI to make judgment calls the way people do. Things like who to marry, whether to seek a new job, what house to buy, how to raise your children, and a myriad of other things. Dr. Michael Gazzaniga, despite being a "brain is a machine" maven, nevertheless points out that humans do not act like machines when questions of subjective judgment are on the line:

Although judges, juries, and attorneys most likely will attribute their stances to various factors, not the least of which are long years of education, philosophical discussion and the like, as usual, most of the goings on in the courtroom are intuitions that we came with from the baby factory, including a sense of fairness, reciprocity, and punishment. Renee Baillargeon and colleagues have been hard at work with a group of toddlers and have shown that a sense of fairness is present not only in two-and-a-half-year-olds, but also sixteen-month-olds. The older group, when asked to distribute treats to animated puppets will do so evenly, and the sixteen-month-old infants prefer cartoon characters that divide prizes equally.

Gazzaniga, Michael S. Who's in Charge? Free Will and the Science of the Brain (p. 231). HarperCollins. Kindle Edition.

A jury, or any other deliberative body, presumably applies a collaborative sense of human-defined justice when considering the case. Could we train AI to do this in a human way? If we placed 12 AI consoles in a jury box, would they, after listening to the prosecution and defense, decide the case in a human way? If they all have the same program, would they always reach unanimous conclusions of guilt or innocence? Or would they debate the way human juries do until they either reach a unanimous consensus of guilty or not guilty or cannot reach a consensus and result in a hung jury?

How would 100 AI consoles debate bills in the Senate and 435 in the House of Representatives? Could they originate legislation based on the current state of information and belief, considered in the context of the Constitution, as our human representatives do? Even if they are not capable of originating legislation, it would still be interesting to see how hundreds of AI consoles would vote on House and Senate bills after hearing the same debates human Representatives and Senators hear. Would they vote yes or no in the same proportions as the real House and Senate? Or would they always reach the same consensus of 100% voting yea or nay? Would they be able to explain the reasons why they voted as they did?

How would AI do on the judiciary? Future President Abraham Lincoln wrote before the Civil War that "the intent of the lawgiver is the law." Attorneys are paid hundreds of dollars per hour to convince judges and juries to interpret the intent of the law in their clients' interest. Differing interpretations of intent and circumstance cause many important Supreme Court decisions to be split f 5-4. The 14th Amendment to the Constitution, passed soon after the Civil War, is currently subject to interpretation of its intent for regulating the citizenship of children born to people who entered the United States illegally:

Fourteenth Amendment

Section 1: *All persons born or naturalized in the United States, and subject to the jurisdiction thereof, are citizens of the United States and of the State wherein they reside. No State shall make or enforce any law which shall abridge the privileges or immunities of citizens of the United States; nor shall any State deprive any person of life, liberty, or property, without due process of law; nor deny to any person within its jurisdiction the equal protection of the laws.*

The courts have previously ruled that children born to persons in the country as tourists, temporary residents on visas, or are in the country illegally are citizens of the United States. However, a person focused on the intent of the 14th Amendment may say it applied only to the children of liberated slaves, to guarantee their rights as U.S. citizens.

I asked ChatGPT: Does the Fourteenth Amendment grant birthright citizenship?

ChatGPT answered:

Yes, the Fourteenth Amendment to the United States Constitution is often interpreted as granting birthright citizenship. The relevant section of the Fourteenth Amendment is the first sentence of Section 1, which states:

"All persons born or naturalized in the United States, and subject to the jurisdiction thereof, are citizens of the United States and of the State wherein they reside."

This clause is commonly referred to as the "Citizenship Clause" and has been the foundation for the principle of birthright citizenship in the United States. It means that anyone born on U.S. soil, with a few exceptions (such as children of foreign diplomats), is automatically granted U.S. citizenship, regardless of their parents' citizenship status.

The Fourteenth Amendment was ratified in 1868, primarily to ensure civil rights and equal protection under the law to former slaves and their descendants after the American Civil War. The Citizenship Clause was intended to overturn the Supreme Court's previous decision in the Dred Scott v. Sandford case, which had denied citizenship to African Americans.

Over the years, the principle of birthright citizenship has been upheld by the U.S. Supreme Court and is considered a fundamental aspect of American citizenship law. However, it is worth noting that the interpretation and application of this principle have been the subject of legal and political debates at various times in U.S. history, with some arguing for more restrictive interpretations or even attempts to amend the Constitution. Nonetheless, as of my last knowledge update in September 2021, birthright citizenship remains a firmly established constitutional principle in the United States.

This answer explains both interpretations fairly and is correct that the current precedent is that birthright citizenship applies to children of parents here illegally.

How would the Supreme Court decide the question now, knowing that the amendment has been repurposed from preventing re-enslavement of children born to African Americans to granting citizenship of children born to parents here illegally? Supreme Court justices who favor immigration under all circumstances will be prone to ruling that children of anyone born here for any reason is an American citizen. Some Justices will not want to overturn the precedent even if they don't agree with those decisions. Other justices may be prone to interpreting its intent as applying only to the descendants of freed American slaves, not to children of foreigners here illegally.

How would an AI program interpret it? Would twelve AI programs rule unanimously based on a common algorithm? Or would the algorithm be randomized to give different responses to refute the notion of robotic conformity? Unless each AI cyber-judge can explain why it ruled one way and rejected counterarguments, I don't believe we could even begin to consider it sentient. Even if we ask these questions in 2123 after a further century of AI development, we may not know whether the AI is sentient in the sense of understanding what it is deciding the way a human being would.

For now, I will only accept being judged by "a jury of my peers" consisting only human beings should the occasion ever arise. I will not consider twelve consoles of AI machines my peers unless and until my brain is copied into a machine and they are judging me as one of their own kind.

About the Author

"Understanding History is a key to understanding the present and extrapolating the future" - Alan Sewell.

I've pursued careers in technology, science, economics, history, and international business, striving to bring clarity to these subjects, by viewing their present states of development in context with their historical origins. Although every day is a new day, the new days are layered on top of repeating cycles of history as old as Mankind.

Regarding American history, I've been commended for "interpreting the American experience" into six critical watershed periods when political and economic crises were resolved to advance the country into the next era of its history:

FRAGMENTATION - FEDERALISM - UNION (1783 - 1815)

SECESSION - WAR - NATIONALISM (1861 - 1868)

WEALTH - DEPRESSION - EXPANSION (1890-1900)

WEALTH - DEPRESSION - LIBERALISM (1925-1933)

CHAOS - HUMILIATION - CONSERVATISM (1968-1981)

INSTABILITY --- RECESSION ---- {Populism or Progressivism} (2008-?)

I've also been commended on writing reviews of books on science and the history of science. I began with the study of science and engineering at Georgia Tech, while writing articles for popular national history magazines, then pursued a career developing algorithms to manage the business of multinational corporations. I have returned to my specialization of presenting history, science, and economics to popular audiences as easily comprehended books and articles, without omitting the core tenets and controversies of the subjects. I am a popular commenter in the Wall Street Journal on-line edition on topics of business, economics, history, and science.

Other Science Books by Alan Sewell:

https://www.amazon.com/dp/B0CC45D4W7

Quantum Shockwave: The Physics and Metaphysics of Quantum Mechanics Kindle Edition
by Alan Sewell (Author)   Format: Kindle Edition
4.0 ★★★★☆ ˅   1 rating   See all formats and editions

Kindle	Paperback
$6.45	$17.65
Read with Our Free App	1 New from $17.65

"Those who are not shocked when they first come across quantum mechanics cannot possibly have understood it," said Niels Bohr, one of the few physicists who ever prevailed in argument with Albert Einstein.

"I like to believe the moon is there even when I'm not looking at it," protested Einstein. "I have thought a hundred times as much about the quantum problem as I have about general relativity theory. I cannot seriously believe in [quantum theory] because ... physics should represent a reality in time and space, free from spooky action at a distance."

"I think I can safely say that nobody understands quantum mechanics," concurred Richard Feynman, a father of nanotechnology and quantum computing.

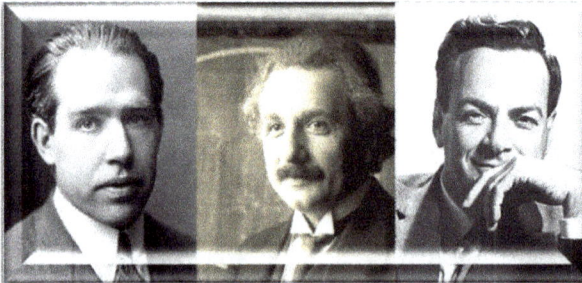

Niels Bohr   Albert Einstein   Richard Feynman

"Those who are not shocked when they first come across quantum mechanics cannot possibly have understood it," said Niels Bohr, one of the few physicists who ever prevailed in argument with Albert Einstein.

"I like to believe the moon is there even when I'm not looking at it," protested Einstein. "I have thought a hundred times as much about the quantum problem as I have about general relativity theory. I cannot seriously believe in [quantum theory] because ...physics should represent a reality in time and space, free from spooky action at a distance."

"I think I can safely say that nobody understands quantum mechanics," concurred Richard Feynman, a father of nanotechnology and quantum computing.

This book presents the often-shocking history, philosophy, controversies, and personalities of quantum mechanics. It explains in popular terms the great debates between Albert Einstein, who was skeptical of quantum theories, and QM's founders, like Niels Bohr. These unresolved controversies are debated as intensively by physicists today. My goal is to condense the meaningful information you might obtain from years of stressful reading of QM books into a day's interesting read. I've referenced QM videos (current through 2023) and recent popular QM books, so you can dig deeper into QM studies if you want, without having to wade through hundreds of books and Youtube videos to find the best ones that are scientifically vetted and interesting for a lay audience.

https://www.amazon.com/dp/B0CBKY7HW1

Einstein's Car: The Engine of Relativity Kindle Edition
by Alan Sewell (Author)   Format: Kindle Edition

See all formats and editions

Kindle	Paperback
$0.00 kindleunlimited	$17.55 prime
Read with Kindle Unlimited to also enjoy access to over 4 million more titles	1 New from $17.55
$4.95 to buy	

This book seeks to provide lay readers the "Eureka!" moment of understanding Relativity without inducing headaches in comprehending the math. It is written for those who have studied it, wondered about it, and still lack an intuitive understanding of it. As well as science, it seeks to open a window into the intellectual spirit of Relativity that animated the minds of its discoverers, including Einstein, Poincare, Minkowski, and Lorentz.

Print length	Language	Sticky notes	Publication date	File size	Page Flip

Albert Einstein and his second wife Elsa

This book seeks to provide lay readers the "Eureka!" moment of understanding Relativity without inducing headaches in comprehending the math. It is written for those who have studied it, wondered about it, and still lack an intuitive understanding of it. As well as science, it seeks to open a window into the intellectual spirit of Relativity that animated the minds of its discoverers, including Einstein, Poincare, Minkowski, and Lorentz.

Reading about Relativity as a non-professional physicist is frustrating. You feel it's an important aspect of the Universe you want to understand. But studying it either balks you with incomprehensible mathematical formulas or tries to placate you with platitudes of "everything is relative except the invariant speed of light" without explaining the how's and why's.

No matter how much information you take in, understanding remains elusive. You find yourself immersed in allegories such as Einstein's Train and Embankment, The Twins Paradox, and even The Andromeda Paradox, explained in contradictory ways. How can we glean an intuitive understanding of how time behaves in Relativity, when in everyday life we only experience time ticking at a constant rate and events pegged to a universally agreed upon timeline?